高等学校工程管理专业规划教材

国际工程管理承包商标准信函

Standard Letters for International Building Contractors

李德智　赵世忠　编著

中国建筑工业出版社

图书在版编目（CIP）数据

国际工程管理承包商标准信函/李德智，赵世忠编著. —北京：中国建筑工业出版社，2017.7
高等学校工程管理专业规划教材
ISBN 978-7-112-20891-3

Ⅰ.①国… Ⅱ.①李… ②赵… Ⅲ.①建筑工程-承包-信函-范文 Ⅳ. ①TU723.1

中国版本图书馆 CIP 数据核字（2017）第 146651 号

考虑到国际工程常用合同文本的特点，结合国际工程实践，本书共分为十章。第一章是国际工程承包合同概述，主要介绍国际工程承包合同的涵义、国际工程的主要参与者、国际工程承包常用合同范本以及承包商的主要权利义务等。通过对这些基础知识的介绍，使合同管理者在起草信函时能够较好地运用相关知识，把握好工程参与者之间的相互关系，使信函写作能够达到预期目的。第二章～第十章，每章分为两个部分，第一部分是对本章内容的中文介绍，第二部分是英文和中文信函。为了使用方便，本书将信函作了适当合理分类，它们分别是投标信函、合同签订与履行前的准备信函、保险信函、现场管理信函、工程款项支付信函、工期延长信函、分包合同信函、工程竣工与合同终止信函、索赔与仲裁信函。

为更好地支持相应课程的教学，我们向采用本书作为教材的教师提供教学课件，有需要者可与出版社联系，邮箱：cabpcm@163.com。

* * *

责任编辑：刘晓翠 张 晶 王 跃
责任校对：李欣慰 党 蕾

高等学校工程管理专业规划教材
国际工程管理承包商标准信函
李德智 赵世忠 编著
*
中国建筑工业出版社出版、发行（北京海淀三里河路 9 号）
各地新华书店、建筑书店经销
北京红光制版公司制版
北京市安泰印刷厂印刷
*
开本：787×1092 毫米 1/16 印张：16¾ 字数：413 千字
2017 年 9 月第一版 2017 年 9 月第一次印刷
定价：**35.00** 元（赠课件）
ISBN 978-7-112-20891-3
（30537）

版权所有 翻印必究
如有印装质量问题，可寄本社退换
（邮政编码 100037）

前　言

在国际工程项目承包过程中，除了个别事项用口头沟通和谈判外，绝大多数项目事宜需要使用正式的书面文件进行沟通，而信函及相关文件的撰写则是书面沟通的主要表现形式。作为工程合同管理的重要手段，从投标阶段开始，承包商就开始通过标准信函与雇主、工程师（建筑师）或其他工程项目相关人员进行沟通，这种沟通形式贯穿工程管理的全过程。因此，良好的函件撰写能力是对合同管理者的一个基本素质要求，也是避免和减少合同管理过程中承包商与相关各方矛盾与误解的有效渠道。

本书的写作过程中，考虑到国际工程常用合同文本的特点，结合国际工程实践，将编写分为十章。第一章是国际工程合同基础知识概述，主要介绍了国际工程承包合同的涵义、国际工程的主要参与者、国际工程承包常用合同范本以及承包商的主要权利义务等。通过对介绍基础知识，使合同管理者能够在起草信函时很好地把握相关知识和工程参与者之间的相互关系，使信函写作能够达到预期目的。第二～第十章每章分为两个部分，第一部分是本章内容的中文介绍，第二部分是信函的英文和中文介绍。为了使用方便，在写作过程中，本书根据国际工程合同管理过程中的不同环节将标准信函作了适当合理分类，它们分别是投标信函、合同签订与履行前的准备信函、保险信函、现场管理信函、工程款项支付信函、工期延长信函、分包合同信函、工程竣工与合同终止信函、索赔与仲裁信函。

本书由沈阳建筑大学商学院李德智与外国语学院赵世忠编写。其中李德智负责本书结构的整体设计、第一章写作以及第二～第十章中文部分写作。赵世忠负责信函的翻译与写作。在本书编写过程中，沈阳建筑大学商学院院长刘亚臣教授、外国语学院院长李家坤教授给予了多方面的指导，并提出了宝贵意见，在此一并表示感谢。

编者

2017 年 3 月

目　录

第一章 国际工程承包合同概述

第一节 国际工程承包合同的主要参与者

国际工程承包是指个人或企业，在国际工程承包市场上通过投标、接受委托或其他途径承揽国际组织、外国政府或私人雇主的工程建设项目、物资采购及其他方面的承包业务，是一种涉及资金、技术、设备、劳务等多方面内容的综合性国际经济合作形式。国际工程承包的主要业务包括：建筑项目的咨询、可行性研究、项目地址选择、勘测和动力的提供；工程施工、设备安装和调试；人员培训；项目建成后的生产组织和指导；专项物资采购和各类经营管理等活动。而国际工程承包合同则是指不同国家的平等自然人、法人、其他组织之间为了实现在某个工程项目的特定目的而签订的设立、变更和终止权利义务关系的法律文件。

国际工程承包合同的主要参与者是指参与合同法律关系依法具有民事权利能力和民事行为能力，享受合同权利、承担合同义务的当事人。只要是具有相应的民事权利能力和民事行为能力的平等的自然人、法人、其他组织就可以签订合同。在国际工程承包合同中，合同主体主要由雇主和承包商构成，同时，与国际工程承包合同相关的主体还包括工程师、分包商、供应商等。

划定和了解国际工程承包合同的主要参与者，有利于信函写作中避免法律关系的错位，写给正确的人才是最有效率的沟通。

一、雇主

雇主（The Employer 或 The Owner）是指工程项目的所有者和拥有者以及其财产的合法继承人。美国 AIA 使用“the Owner”或者“Building Owner”代表雇主的含义；而在其他英联邦国家的建筑工程领域，雇主被称为“the Employer”。在标准合同格式中，如 FIDIC、NEC、JCT、ICE 等均使用“the Employer”。

雇主可以是私人、公司，也可以是政府部门或其他社会组织。在国际工程建设领域，绝大多数雇主是公司和政府部门。FIDIC《施工合同条件》（1999 年“新红皮书”）对雇主的解释为：“雇主（Employer）指在投标函附录中指定为雇主的当事人或此当事人的合法继承人”。在 ICE 合同条款中，雇主（Employer）被定义为在投标函附录第一部分中命名为雇主的个人、公司或其他组织，并且包括雇主的代表、继承人和合法受让人。在美国 AIA A201-2007 中，“Owner”是指合同中确定的单数的，体现在所有合同文件中的个人或实体。

二、承包商

承包商是指与雇主签订工程合同，负责实施、完成和维护工程项目的当事人。按照 FIDIC《土木工程施工合同条件》（1987 年第四版）第 1.1（a）（ii）款的规定，承包商（the Contractor）是指“其投标已为雇主接受的当事人以及取得此当事人资格的合法继承人，但除非雇主同意，不指此当事人的任何受让人。”FIDIC 建筑施工合同条款（1999版，以下简称“新红皮书”）第 1.1.2.3 款规定承包商的定义是：“承包商是指已为雇主接受的投标函中指明作为承包商的当事人及其合法继承人。”在美国 AIA A201-2007 Clause3.1.1 中承包商（Contractor）是指合同中确定的单数的，体现在所有合同文件中的个人或实体。根据项目所在地法律要求，承包商需取得合法资格。承包商的主要义务就是在合同规定的时间内和场地内实施和完成他所签约的工程，如工程有缺陷，承包商有义务在缺陷责任期内修补缺陷。大致总结，与承包商相关的合同关系主要有：

1. 分包合同

承包商在承包合同下可能订立许多分包合同，而每个分包商仅完成总承包商的部分工程，向总承包商负责。总承包商向雇主承担全部工程责任，负责工程的管理和所属各分包商工作之间的协调，以及各分包商之间合同责任界限的划分，同时承担协调失误造成损失的责任，向雇主承担工程风险。

2. 供应合同

承包商为工程所进行的必要的材料和设备的采购和供应，与供应商签订供应合同。

3. 运输合同

承包商为解决材料和设备的运输问题而与运输、快递公司签订的合同。

4. 加工承揽合同

即承包商将建筑构配件、特殊构件加工任务委托给加工承揽公司而签订的合同。

5. 租赁合同

在建设工程中承包商常需要施工设备、运输设备、周转材料。当有些设备、周转材料在现场使用率较低，或自己购置需要大量资金投入而自己又不具备相应的经济实力时，可以采用租赁方式，签订租赁合同。

6. 劳务供应合同

即承包商与劳务供应商之间签订的合同，由劳务供应商向工程项目提供劳务。

7. 保险合同

承包商按合同要求对工程财产、人员等进行保险，与保险公司签订保险合同。

这样总承包合同就由不同层次、不同种类的合同组成，这些合同之间存在着复杂的内部联系，它们共同构成该国际工程的合同体系。其中总承包合同是最有代表性、最普遍，也是最复杂的合同类型。它在工程项目的合同体系中处于主导地位，是整个工程项目合同管理的重点。无论是雇主、工程师或承包商都将它作为合同管理的主要对象。

三、工程师

在国际工程合同施工领域，FIDIC 称工程师为 the Engineer，AIA 和英国 JCT 等标准合同文本称工程师为 Architect，也叫建筑师，虽然名称不同，但其职责范围大同小异。鉴于“the Engineer”应用得较广，本书使用 Engineer 代表工程师（建筑师）的含义。工程师又称监理工程师或咨询工程师，是指由雇主聘任代表雇主对工程项目质量、进度、工艺和成本等进行监督管理的人。在 FIDIC“红皮书”等合同条件下，雇主可能委托工程师完成对工程项目的设计任务，并授权其对工程项目施工进行监理。

FIDIC“新红皮书”第 1.1.2.4 款规定工程师的定义是：“工程师指雇主为合同目的而指定作为工程师并在招标附录中保持这一称谓的人员；或者雇主根据第 3.4 条随时指定

的并通知承包商的任何其他人员。”工程师不属于雇主和承包商之间合同关系的一方，按照建筑和土木工程业界惯例，雇主和工程师之间将签订咨询服务合同，明确雇主和工程师的权利和义务及其权利的限制。雇主和工程师之间的服务合同可参考FIDIC出版的《雇主和咨询工程师之间协议书国际通用规则》的格式或其他标准合同格式。

工程师作为代表雇主进行工程项目管理，必须履行合同规定的职责。工程师可以行使合同中规定的或合同必然暗示的权力，但是，如果雇主任命工程师的条款规定，要求工程师在行使权力前必须取得雇主的批准，则此要求应详细列入合同中，进而认为，工程师所行使的权力都已取得了雇主必要的批准。除在合同中有明确的规定外，工程师也无权修改合同或解除合同规定应由承包人所尽的义务。

1999版“新红皮书”已经将工程师定义为雇主人员，雇主与工程师之间的问题相当于雇主内部问题。所以为了合同目的，每当工程师履行或行使合同规定或隐含的义务或权力时，应视为代表雇主。

四、分包商

FIDIC分包合同（1994年版）第4.2款规定：“分包商应承担并履行与分包工程有关的主合同规定承包商的所有义务和责任。”分包商与主承包商相同义务原则产生于主承包商在主合同项下义务的传递或平行转移，这种传递或平行转移有合同上的原因：

1. 主承包商合同义务的来源

主承包商合同义务来源于雇主和主承包商之间签订的主合同，它与雇主签订了主合同后，享有合同权利，承担合同义务。除主合同规定的权利义务外，主承包商还享有法律规定的权利，承担法律规定的义务。

2. 分包商合同义务的来源

尽管分包合同仅仅是分包商与主承包商之间签订的合同，与雇主与主承包商之间签订的主合同没有直接关系，但分包合同是从属性合同，它以主合同的存在为前提，随主合同消灭而消灭。分包商的义务来源于分包合同，但却以主合同中主承包商的义务为最终源头。

3. 分包工程是主总承包工程的一部分

分包商实施的分包工程不是独立于主合同之外的工程，而是构成主合同工程的一部分工程内容。对于主合同工程，主承包商负有全面履行之责，而对于分包工程，分包商也应像主承包商一样履行分包工程。

第二节 国际工程承包常用合同范本

一、国际工程承包常用合同范本概述

从19世纪开始，由于国际工程建设的迅猛发展，英国和爱尔兰就有了自己的建筑工程施工合同标准文本。在英国皇家建筑师协会（RIB）的倡导下，建筑工程施工合同范本得以使用，该范本后来被称为RIBA范本。英国土木工程协会（ICE）创建于1818年，该协会是代表土木工程师的专业机构及资质评定组织，制定了适用于土木工程建设的ICE合同系列。1945年12月，ICE编制出版了第一版ICE合同范本，1950年1月出版了第二版ICE合同，1951年3月、1955年4月、1973年6月、1991年1月分别编制出版了第三版、第四版、第五版、第六版ICE合同范本，目前正在使用的版本是1999年发布的第七版《ICE合同（计量版）》（The ICE Conditions of Contract，Measurement Version，7th edition）。ICE标准合同后来成为FIDIC标准合同范本制定的参考。英国合同审定联合会（JCT）创建于1931年，自该协会成立之日起，就着手制定了JCT合同范本，其合同主要版本有1939、1963、1980、1998、2005和2011年版。目前JCT合同范本是英国所有合同范本中使用范围最广，使用人数最多的合同范本。FIDIC是指国际咨询工程师联合会，由三个欧洲法语国家创立于1913年。FIDIC的成员来自于全球各地近100个国家和地区，中国1996年正式成为FIDIC会员。FIDIC编制了许多标准合同范本，并被广泛使用于世界各地。

本书的承包商信函是以以下合同文本为基础的。在信函写作过程中，承包商应深入研究工程适用文本的基本内容和合同文本的配套文件；同时研究国际工程某些问题条款的国际实践和通常解决办法。在此基础上形成的信函才能有的放矢，达到预期效果。

二、FIDIC合同范本

FIDIC是国际咨询工程师联合会（Fédération Internationale Des Ingénieurs Conseils）的法文缩写。FIDIC的本义是指国际咨询工程师联合会组织。

国际咨询工程师联合会是1913年由欧洲三国独立的咨询工程师协会在比利时根特发起成立。目前有70多个成员国加入，分属于四个地区性组织，即亚洲及太平洋地区会员协会（ASPAC）、欧洲共同体会员协会（CEDIC）、非洲会员协会组织（CAMA）、北欧会员协会组织（RINORD）。FIDIC是最具有权威性的咨询工程师组织，自建立以来，有效地推动了全球范围内高质量的工程咨询服务业的发展。FIDIC总部设在瑞士洛桑，主要职能机构有：执行委员会、土木工程合同委员会、雇主与咨询工程师关系委员会、职业责任委员会以及秘书处。1996年10月中国工程咨询协会正式加入FIDIC，取得了在FIDIC的发言权和表决权。这增加了中国企业开展国际交流、了解国外信息和开拓对外业务的机会。

FIDIC成立100多年来，对管理建设施工项目、促进国际经济技术合作的发展起到了重要作用。由该组织编制的《土木工程施工合同条款》（“红皮书”）、《电气与机械工程合同条款》（“黄皮书”）等合同条款被世界银行、亚洲开发银行等国际和区域发展援助金融机构作为实施项目的合同协议范本。这些合同协议范本，内容严密，对履约各方和实施人员的职责义务作了明确的规定；对实施项目过程中可能出现的问题也都有较为合理的规定和解决办法。这些协议性文件为科学管理实施项目提供了可靠的依据，有利于保证工程质量、工期和控制成本，使雇主、承包商以及咨询工程师等有关人员的合法权益得到尊重。

FIDIC系列合同范本经历了漫长的发展过程。从其1957年颁布第一版土木工程施工合同条件以来，到1999年以前，该组织共制定和颁布了《土木工程施工合同条款》、《电气与机械工程施工合同条款》、《客户/咨询工程师（单位）服务协议书范本》，《设计—建造与交钥匙工程合同条款》、《工程施工分包合同条款》等合同系列。1999年FIDIC又将这些合同体系作了重大修改，颁布了一套全新的标准合同范本，包括《建筑施工合同条款》（“新红皮书”）、《生产设备与设计—建造合同》（“新黄皮书”）、《设计采购施工（EPC）/交钥匙工程合同条款》（“银皮书”）和适合于小规模项目的《简明合同格式》（“绿皮书”）。

合同范本是FIDIC组织制定的适用于建设工程施工的标准合同范本，主要包括以下几种：

(1)《建筑工程合同条款》(Conditions of Contract for Construction)（“新红皮书”)。

该合同范本适用于由雇主负责设计的建筑施工工程，其最大特点有以下两个方面，其一是雇主支付承包商的依据是工程量清单；其二，工程师的纠纷解决职能被DAB机构所取代。

(2)《生产设备与设计建造合同条款》(Conditions of Contract for Plant and Design-Build for Electrical and Mechanical Plant，and for Building and Engineering Works，Designed by the Contractor)（“新黄皮书”)。

“新黄皮书”适用于大型工程的设备提供和施工安装，承包工作范围包括设备的制造、运送、安装和保修几个阶段。这个合同是在土木工程施工合同基础上编制的，针对相同情况制定的条款完全照搬土木工程施工合同的规定。与“新红皮书”的区别主要表现为：一是该合同涉及的不确定风险的因素较少，但实施阶段管理程序较为复杂；二是工程设计任务由承包商承担。

(3)《设计采购施工（EPC）/交钥匙项目合同条款》(Conditions of Contract for EPC Turnkey Projects，First Edition，1999。该文件适用于以交钥匙方式提供工厂或类似设施的加工或动力设备（“银皮书”)、基础设施项目或其他类型的开发项目，采用总价合同。该合同范本适用于建设项目规模大、复杂程度高、承包商提供设计、承包商承担绝大部分风险的情况。

(4)《简明格式合同（Short Form of Contract)》(“绿皮书”)。“绿皮书”的宗旨在于使该合同范本适用于投资规模相对较小的民用土木工程，如：①造价在500000美元以下以及工期在6个月以下；②工程相对简单，不需专业分包合同；③重复性工作；④施工周期短。

(5) MDB（Multilateral Development Banks）(“粉皮书”)是“红皮书”的harmonised edition，初版发布于2005年5月，修订于2006年3月和2010年6月。“粉皮书”(Pink book）是FIDIC与世界银行等国际金融组织合作专门编制的用于国际多边金融组织出资的建设项目，其条款并不是很适合其他资金来源的项目，例如，MDB版本对采购地就有明确而严格的限制。

(6)《设计、建造和项目运营合同条款》(Conditions of Contract for Design，Build and Operate Projects)（“金皮书”)。“金皮书”将设计、建设、运营和维护于一体，项目的运行试验要经受20年运营与维护的考验，在此期间，承包商要实现运作目标，并根据合同条件，将项目的运营权转给雇主。FIDIC声明“金皮书”的起草的目的在于减少项目移交后由于项目设计、工艺和材料质量问题引起的风险。因此，“金皮书”适用于需要长时间使用的，并需要长期维护的基础设施项目，例如PPP和BOT建设项目。

(7)《设计—建造与交钥匙工程合同条款》（“橘皮书”）(Conditions of Contract for Design-Build and Turnkey，1st Edition 1995)。

“橘皮书”发布于1995年，为FIDIC合同的工作小组起草合同提供样本。在1999年FIDIC发布了“新红皮书”和“新黄皮书”后，就很少使用“橘皮书”了。

(8)《招标程序》(Tendering Procedure）（“蓝皮书”)。国际咨询工程师联合会在

1982年出版了《招标程序》，反映了国际上建设行业当今招标投标的通行做法，它提供了一个完整、系统的国际建设项目招标程序，具有实用性、灵活性。

(9)《雇主与咨询工程师标准服务协议书》(Conditions of the Client/Consultant Model Services Agreement)("白皮书")。"白皮书"是雇主为了工程项目的顺利进行与咨询机构签订的关于工程咨询服务的合同。

(10)《土木工程施工分包合同》。FIDIC编制的《土木工程施工分包合同》是与FIDIC施工合同条款配套使用的分包合同范本。FIDIC分包合同制定于1994年，2011年进行了修订。

三、英国制定的建筑工程施工合同范本

英国建筑业编制工程合同范本的机构主要有两个，一是英国土木工程师学会(Institution of Civil Engineer，ICE)，另一个是英国合同审定联合会(Joint Contracts Tribunal，JCT)。

(一) ICE

英国土木工程师学会创建于1818年，是英国代表土木工程师的专业机构及资质评定组织，在国际上也有广泛影响。ICE的成员包括从专业土木工程师到学生在内的会员近8万名，其中五分之一分布在英国以外的150多个国家和地区。ICE是根据英国法律具有注册资格的教育、学术研究和资质评定的团体，其出版的合同范本成为国际知名范本之一。

ICE正式用于施工的合同条件第一版于1945年出版，其后陆续进行了修改，目前最新版本是1999年第七版。现行ICE系列合同条件如下：

(1)《ICE合同条件，工程量计量模式》，第七版(ICE Conditions of Contract，Measurement Version，7th Edition)。

(2)《ICE合同条件，定期模式》，第一版(ICE Conditions of Contract，Term Version，lst Edition)。

(3)《ICE小型工程合同条件》，第三版(ICE Conditions of Contract for Minor Works，3rd Edition)。

(4)《ICE合同条件，伙伴关系补遗》，第一版(ICE Conditions of Contract，Partnering Addendum，lst Edition)。

(5)《ICE合同条件，考古勘察版》，第一版(ICE Conditions of Contract，Archaeological Investigation，lst Edition)

(6)《ICE合同条件，地质勘察版》，第二版(ICE Conditions of Contract，Ground Investigation，2nd Edition)。

(7)《ICE合同条件，目标成本模式》，第一版(ICE Conditions of Con-tract，Target Cost Version，lst Edition)。

多年来，ICE编制的合同范本在世界各地得到了广泛采用，其中影响最大的是应用于土木工程施工的《ICE合同条件，工程量计量模式》。早期的FIDIC合同条件，如"红皮书"第四版及以前的版本主要是基于该ICE合同条款的框架制定的。ICE也专门为分包工程、设计—建造模式工程制定了合同范本。

(二) NEC

除了上述传统的合同条款外，ICE于20世纪90年代初研究制定了一套新型的合同范本，称为“新工程合同”（New Engineering Contract，以下简称 NEC）。

NEC 第一版于 1993 年 3 月出版，第二版于 1995 年出版，更名为“工程施工合同（ECC）”。之后逐渐形成了 NEC 合同系列合同范本。

最新版的 NEC 系列合同范本是 2005 年 7 月出版的第三版，简称 NEC3。NEC3 包含以下六类合同文件：

（1）工程施工合同（Engineering and Construction Contract，ECC），也称“黑皮书”，适用于所有工程领域的土木工程施工，包括设计工作。

（2）专业服务合同（Professional Services Contract，PSC），用于聘用专业咨询人员、项目经理、设计师、监理者等专业技术人员或机构。

（3）工程施工简明合同（Engineering and Construction Short Contract，ECSC），适用于结构简单，风险较低，对项目管理要求不太苛刻的工程项目。

（4）合同争议评判员合同（Adjudicator's Contract），雇主聘任“合同争议判员”的合同。

（5）定期服务合同（Term Service Contract，TSC），用于采购有固定期限的服务。

（6）框架合同（Framework Contract，FC），这是 NEC3 新增加的合同范本，用于在雇主和承包商之间完全确定项目内容之前建立的一种工作关系。

NEC 合同范本还特别指出，成功使用 NEC3 的关键是一种“文化转变”，即把传统的工程合同关系从一种被动的管理与决策模式转变为有远见的创造性的合作关系，简言之，就是从对抗型的项目组织模式转变成为合作型的项目组织模式。

（三）JCT

除上述两个建设工程标准合同体系外，英国合同审定联合会（Joint Contracts Tribunal，JCT，创建于 1931 年）组织制定了 JCT 合同体系。目前，JCT 合同体系包括 1998 年版标准格式合同和 2005 年版合同族。在 2005 年合同族中，每一种合同类型均包括主合同和分包合同标准文本，以及其他能够跨越不同合同版本的标准文件。2005 年版合同族的主要合同文件类型及其主合同文件如下：

（1）小型工程建设合同（Minor Works Building Contract）；

（2）中型工程建设合同（Intermediate Building Contract）；

（3）标准建造合同（Standard Building Contract，分为有工程量清单、有估计工程量、无工程量清单三种）；

（4）设计建造合同（Design and Build Contract）；

（5）大型工程施工合同（Major Project Construction Contract）；

（6）JCT 构建卓越合同（JCT-Constructing Excellence Contract）；

（7）施工管理专业合同（Construction Management Trade Contract）；

（8）建造管理合同（Management Building Contract）；

（9）住宅许可工程建造合同（Housing Grant Works Building Contract）；

（10）测量期合同（Measured Term Contract）；

（11）主要成本建造合同（Prime Cost Building Contract）；

（12）修复和维护合同（Repair and Maintenance Contract）；

（13）框架协议（Framework Agreement）；

(14) 仲裁协议（Adjudication Agreement）。

JCT 和 ICE 是英国建筑业使用最为广泛的合同文本，但 JCT 合同更常用于房屋建设工程项目中。

四、美国制定的建设工程施工合同范本

美国建筑业编制合同范本的机构主要有以下几个：

(一) AIA 及其合同范本

AIA 创始于 1857 年，是美国主要的建筑师专业社团。其制定并公布了一系列的标准合同范本，在美国建筑业界及国际工程承包界特别是在美洲地区具有较高的权威性。

1911 年 AIA 首次出版了《施工通用条件》(General Conditions for Construction)。经过多年的发展，AIA 形成了一个包括 90 多个独立文件在内的复杂体系。和英国合同范本相比，AIA 合同范本的主要特点是为各种工程项目管理模式制定了不同的协议书，而同时把通用条件作为独立文件单独出版。

此外，AIA 还随时关注建筑业的发展趋势，每年都对部分文件进行修订或重新编写。例如，2004 年更新了 12 份文件，2005 年更新了 6 份文件，并会每隔十年左右对合同体系及内容进行较大地调整。特别是在 2007 年，AIA 对整个文件体系的编号系统及内容都进行了较大规模地调整。

2007 年修改后的 AIA 系列合同范本根据文件性质的不同分为 A、B、C、D、E、G 六个系列：

A 系列：雇主与总承包商、CM 经理、供应商之间，总承包商与分包商之间的合同文件（协议书及合同条件）；施工合同通用条件以及与招标投标有关的文件，如承包商资格申报表，各种保证的标准格式；

B 系列：雇主与建筑师之间的合同文件；

C 系列：建筑师与专业咨询机构之间的合同文件；

D 系列：建筑师行业有关文件；

E 系列：电子文件协议附件；

G 系列：合同和办公管理中使用的文件。

在 AIA 系列合同条件中，AIA A201-2007 是其专用的施工合同通用条件，与 1997 年的版本相比作了较多修改。

(二) AGC 及其合同范本

AGC 是 Associated General Contractors 的简写，是美国总承包商协会。该协会成立于 1918 年，是美国最大的、历史最悠久的建筑行业协会组织，由有资格的建筑承包商和与建筑业相关的公司组成，代表承包商发出来自建筑业的声音，在美国经济发展中发挥着很重要的作用。AGC 作为承包商的行业组织，独立制定了一套与 AIA 文件功能相近的标准合同文件范本，和 AIA 出版的合同文件相比，AGC 更能照顾到承包商的利益。

除了独立编制自己的 AGC 合同范本外，AGC 还积极参与 AIA 等其他组织主持编制的合同范本。随着美国建筑业的发展，2008 年初，AGC 联合代表雇主的行业组织等，对 AGC 现有的合同进行了全面地修订，并将修订后的版本称为“合议文件”（Consensus Docs），得到了 28 个与工程建设业相关的主要协会的认可。

合议文件系列包括 90 多个合同范本文件，覆盖了所有工程建设模式下的合同文件需求，包括六大系列：200系列、300系列、400系列、500系列、700系列、800系列。

（三）EJCDC及其合同范本

EJCDC是Engineers Joint Contract Documents Committee的简写，是美国工程师联合合同文件委员会。该委员会（EJCDC）是一个以工程师为主要代表又兼顾其他机构的合同委员会。这个机构旨在编制公平、客观的合同范本，并能反映出工程建设中的最新管理思想。EJCDC编制的合同范本类型包括：

(1) 施工（Construction）；

(2) 设计—建造（Design-Build）；

(3) 环境整治（Environmental Re-mediation）；

(4) 联营体、同行评议及其他协议（Joint Venture，Peer Review，and Other Agreements）；

(5) 雇主与工程师（Owner & Engineer）；

(6) 工程师与咨询分包商（Engineer& Subconsultant）；

(7) 贷款机构合同版本（Funding Agency Editions）；

(8) 采购（Procurement）。

除设计—建造的合同范本文件是2009年版外，其他最新的合同范本都为2007年版。整个范本系列中最基本的部分是用于工程施工的合同范本，称为C系列。整套施工合同范本的核心是通用条件（C-700），一般与相应的雇主与承包商协议书配合使用，共有两种雇主与承包商协议书可供选用，分别用于总价合同和成本补偿合同。C系列还包括一整套供雇主用于招标投标的文件，包括投标书格式、各种保函格式等。

第三节 国际工程承包商的主要权利和义务

一、承包商的主要权利和义务概述

承包商的权利与义务是国际工程承包合同的核心内容。承包商的权利对应的是雇主相应的义务；承包商的义务，则对应雇主相应的权利。例如，承包商取得工程款项的权利，对应的是雇主付款的义务；而承包商对指定分包商的不接受权则对应雇主的分包商提名或指定的权利。因此，承包商在签订承包合同前后应了解和理解承包合同中的权利和义务，在此基础上，积极与雇主以及工程师进行沟通，使权利义务能够顺利地履行和落实，避免权利义务之间的失衡引起纠纷。这些纠纷既不利于工程的顺利进行，也会给承包商增加不必要的费用。

承包商信函是以承包商为主体发出的意在解决工程施工过程中各种问题的工程管理方式之一。因此，承包商首先要充分了解自身的权利义务，这样才能更好地保护自身权利，避免不利于已的情况发生。

根据FIDIC、AIA、JCT等国际工程合同范本，承包商的主要权利和义务包括以下内容。

二、承包商的主要权利

（一）进入现场的权利

在FIDIC合同条款及附录中，有关进入现场的文件为承包商进入现场提供方便，如果合同条款及附录中没有相关文件，雇主有义务根据项目要求为承包商进入和占有现场提供方便。在这种情况下，慎重的承包商在雇主进行现场移交文件时，要做好各项准备工作，全部或部分移交不能迟于中标通知书发出的42天内。

现场移交对于FIDIC合同执行来说是非常重要的，应当进行适当的管理。移交是雇主的责任，而非工程师。雇主移交了现场，承包商人员才能进入，才能准备开工，因此，承包商应认真检查现场，调查非承包商责任引起的潜在的障碍。同时，也要及时检查现场的基础设施，例如水、电等配套设施，为开工做准备。

（二）请求雇主付款的权利

根据各国的实践和相关规定，承包商是指有一定生产能力、技术装备、流动资金，具有承包工程建设任务的营业资格，在建筑市场中能够按照雇主的要求，提供不同形态的建筑产品，并获得工程价款的建筑企业。承包商交给雇主的是建筑产品——完工的、符合质量要求的建筑物；雇主则会根据工程的进度向承包商支付工程价款，包括工程成本、税金和利润。例如美国AIA A201进行了为期10年的修订工作（1997～2007年），修改后的条款第2条规定，承包商在怀疑雇主支付能力的情况下，可以要求雇主提供合理的资金安排计划。又如，FIDIC“新红皮书”（1999版）第14条赋予了承包商向雇主索要工程款的权利。

（三）索赔权利

索赔是承包商弥补损失、赢取利润的重要手段之一。FIDIC“新红皮书”第20.1款规定：如果承包商认为，根据本条件任何条款或与合同有关的其他文件，他有权得到竣工时间的任何延长期和（或）任何追加付款，承包商应向工程师发出通知，说明引起索赔的事件或情况。

（四）指定分包商不接受权

“新红皮书”（1999版）第5.2款规定了承包商对指定分包商的不接受权，如果1）有理由相信，该分包商没有足够的能力、资源或财力；2）分包合同没有明确规定，指定的分包商应保障承包商不承担指定的分包商及其代理人和雇员疏忽或误用货物的责任；3）分包合同没有明确规定，对分包的工作（包括设计，如果有），指定的分包商应：1）为承包商承担此项义务和责任，能使承包商履行其根据合同规定的义务和责任，和2）保障承包商免除根据合同规定或与其有关产生的所有义务和责任，以及因分包商任何未能完成这些义务或履行这些责任的影响产生的义务和责任。

（五）终止合同权

FIDIC“新红皮书”（1999版）第16.2款规定：如出现下列情况，承包商应有权终止合同：

(1) 承包商在根据第16.1款［承包商暂停工作的权利］的规定，就未能遵循第2.4款［雇主的资金安排］规定的事项发出通知后42天内，仍未收到合理的证据；

(2) 工程师未能在收到报表和证明文件后56天内发出有关的付款证书；

(3) 在第14.7款［付款］规定的付款时间到期后42天内，承包商仍未收到根据期中付款证书的应付款额（按照第2.5款［雇主的索赔］规定的扣减部分除外）；

(4) 雇主实质上未能根据合同规定履行其义务；

(5) 雇主未遵守第1.6款［合同协议书］或第1.7款［权益转让］的规定；

（6）第 8.11 款［拖长的暂停］所述的拖长的停工影响了整个工程；或

（7）雇主破产或无力偿债，停业清理，已有对其财产的接管令或管理令，与债权人达成和解，或为其债权人的利益在财产接管人、受托人或管理人的监督下营业，或采取了任何行动或发生任何事件（根据有关适用法律）具有与前述行动或事件相似的效果。

（六）暂停工作的权利

"新红皮书"（1999 版）第 16.1 款规定：如果工程师未能按照第 14.6 款［期中付款证书的颁发］的规定确认发证，或雇主未能遵守第 2.4 款［雇主的资金安排］或第 14.7 款［付款］的规定，承包商可在不少于 21 天前通知雇主，暂停工作（或放慢工作速度），除非并直到承包商根据情况和通知中所述，收到了付款证书、合理的证明或付款为止。

（七）警告权利

"新红皮书"（1999 版）第 1.8 款规定："如果一方发现为实施工程准备的文件中有技术性错误或缺陷，应迅速将该错误或缺陷通知另一方。""黄皮书"和"银皮书"（1999 版）第 1.8 款也规定了相同的内容。

FIDIC 分包合同（1994 版）第 2.1 条规定："分包商在审阅分包合同和（或）主合同时，或在分包合同工程的施工中，如果发现分包工程的设计或规范存在任何错误、遗漏、失误或其他缺陷，应立即通知承包商。"一般而言，无论分包合同中是否明示规定此类警告义务，法律默示分包商对设计错误等负有警告义务。

FIDIC 分包合同将分包商发现设计或规范存在错误、遗漏、失误或其他缺陷的警告义务的默示义务上升为明示条款，这是 FIDIC 分包合同对 1987 年"红皮书"（第四版）的一项重要补充。

三、承包商的主要义务

（一）承包商在投标阶段的义务

投标是承包商获得工程订单的关键一步，因此，在制定投标策略，准备投标文件时，承包商应检查工地现场和周围的环境。特别是对工地现场地下的地理状况有所了解，收集和分析任何与施工地有关的信息，包括但不限于水文信息、地下和地表结构、气候条件等。一旦中标，在 FIDIC 和 AIA 合同条件下，地质的好与坏都将由承包商承担责任；同时承包商应注意获取市场信息，注重开发和调研。在投标阶段要认真细致地调查市场情况，研究与招标文件有关的各种资料以及现场情况，使自己的投标基于在投标报价范围内能够完成项目任务的基础上。在如愿中标之后，应谨慎准备与雇主的谈判，制定谈判策略，根据项目实际情况争取于己有利的方案。

（二）中标后到开工前承包商的主要义务

中标后，承包商要积极地与雇主商谈合同细节，补充合同文件，尽快完成合同的签订工作。这样，信函往来则成为完善合同条款的重要手段。合同签订后，承包商筹备组织相关人力、设备、劳务等资源，配合雇主单位办理与施工有关的各种许可；积极与工程师沟通，报送施工方案、项目经理组成和专项方案；搭设施工现场临时设施、准备施工用水、电及场内道路，布设各种制作、加工场地；与雇主单位进行平面坐标点的书面交接；进行图纸会审等。

（三）实施和完成工程项目的义务

实施和完成工程项目是承包商的一项最基本的、最重要的义务。FIDIC"新红皮书"

规定了承包商的一般义务，概括如下：（1）承包商的主要义务。承包商的主要义务是根据合同的规定和工程师的指示，实施和完成工程，并修补工程中的任何缺陷。（2）提供设备的义务。承包商应提供合同规定的生产设备、承包商人员、货物、消耗品或其他物品和服务。（3）现场施工的义务。承包商应对所有现场作业、所有施工方法和全部工程的完备性、稳定性和安全性承担责任。（4）提供施工方法的义务。在工程师要求时，承包商应提交其建议采用的工程施工安排和方法的细节。如事先未通知工程师，对这些安排和方法不得作重要的改变。（5）设计义务。承包商应在合同规定的范围内进行设计工作。如果承包商承担了设计工作，设计应满足使用功能。

（四）质量义务

承包商的质量义务就是承包商对承包的工程质量负责。质量义务分为明示的质量义务和默示的质量义务。明示的质量义务，即承包商实施的工程应符合合同文件的要求。在“新红皮书”中，明示质量义务指工程质量应符合设计文件、规范的要求。同时，在施工合同中，默示质量义务主要指雇主对包括设计、材料、工艺、建筑师或工程师的满意。为了保证工程项目的质量，“新红皮书”在多处条款中对质量问题作了规定，主要包括：

1. 承包商应建立质量保证体系。该体系应符合合同的详细规定。工程师有权对质量体系的任何方面进行审查。具体而言，承包商应在每一设计和实施阶段开始前，向工程师提交所有程序和如何贯彻要求的文件的细节，供其参考。当任何技术性文件发给工程师时，文件本身应有经承包商本人事先批准的明显证据。遵守质量保证体系，不应解除合同规定的承包商的任何义务或职责。

2. 雇主人员应在所有合理的时间内：（1）有充分机会进入现场的所有部分以及获得天然材料的所有地点；（2）有权在生产、加工和施工期间（在现场和其他地方），检查、检验、测量和试验所用材料和工艺，检查生产设备的制造和材料生产加工的进度。承包商应为雇主人员进行上述活动提供一切机会，包括提供进入条件、设施、许可和安全装备。此类活动不应解除承包商的任何义务或职责。

3. 修补工作。尽管已有先前的任何试验或证书，工程师仍可指示承包商进行以下工作：（1）将不符合合同要求的任何生产设备或材料移出现场，并进行更换；（2）去除不符合合同的任何其他工作，并重新实施；（3）实施因意外、不可预见的事件或其他原因引起的、为工程的安全迫切需要的任何工作。承包商应在指示规定的合理时间（如果有）内执行该指示，或在上述（3）项规定的紧急情况下立即实施。

4. 完成扫尾工作和修补缺陷。为了使工程、承包商文件和每个单位工程在相应缺陷通知期限期满日期或其后，尽快达到合同要求（合理的损耗除外），承包商应：（1）在工程师指示的合理时间内，完成接收证书注明日期时尚未完成的任何工作；（2）在工程或单位工程（视情况而定）的缺陷通知期限期满日期或其以前，按照雇主（或其代表）可能通知的要求，完成修补缺陷或损害所需的所有工作。如果出现缺陷或发生损害，雇主（或其代表）应相应地通知承包商。

（五）进度义务

承包商或分包商应在合同规定的工期内完成合同规定的工程项目，这是承包商或分包商的一项基本义务。大多数建筑工程承包标准格式合同在通用条款、专用条款、合同协议书以及投标附录中均规定了具体的工期，但同时，在 FIDIC、JCT、ICE 等标准格式合同的通用条款中，也规定了以“应有的速度和毫不拖延地（Due Expedition and Without Delay)”或者“正常地和勤勉地（Regularly and Diligently)”进行施工表示承包商或分包商的时间义务的明示条款。“新红皮书”（1999版)、“黄皮书”和“银皮书”第8.1款规定：“承包商应在开工日期后，在合理可能的情况下尽早开始工程的实施，随后以应有的速度并毫不耽搁地进行工程。”承包商的时间义务主要是由施工合同或分包合同中的明示条款和默示条款规定。施工合同以及分包合同一般均明确规定承包商的明示义务，在规定的日期或在规定的时间之内完成工程。

（六）合作义务

FIDIC“新红皮书”（1999版)、“新黄皮书”和“银皮书”第4.6款规定了合作义务。在施工合同中，承包商的合同义务包括：(1) 及时提供报表、文件、索赔通知；(2) 向建筑师/工程师提交质量保证体系文件；(3) 提交进度计划、图纸、规范和其他文件；(4) 避免干扰；(5) 现场保安；(6) 安全保障；(7) 环境保护；(8) 保护健康。

（七）设计义务

FIDIC“新红皮书”（1999版）和2005年协调版施工合同中，承包商是否承担设计义务，应视合同的具体规定确定。在雇主负责提供设计时，承包商的主要义务是按照雇主提供的设计进行施工，完成和修补其中的缺陷工程。如果合同要求承包商负责设计，则承包商应在合同规定的范围进行设计，并对其设计承担全部的责任。FIDIC“新红皮书”、2005年协调版施工合同第4.1款规定了承包商的设计应满足使用功能的义务。

（八）提供保证、保障和保险的义务

FIDIC“新红皮书”（1999版）中，承包商也承担了提供保证、保障和保险的义务，主要包括：(1) 承包商应对严格履约（自费）取得履约担保，保证金额和币种应符合投标书附录中的规定确定。如投标书附录中没有提出保证金额，本款应不适用。承包商应在收到中标函后28天内向雇主提交履约担保，并向工程师送一份副本。履约担保应由雇主批准的国家（或其他司法管辖区）内的实体提供，并采用专用条件所附格式或雇主批准的其他格式。承包商应确保履约担保直到其完成工程的施工、竣工及修补完任何缺陷前持续有效和可执行。(2) 承包商应保障和保持使雇主、雇主人员以及他们各自的代理人免受以下所有索赔、损害赔偿费，损失和开支（包括法律费用和开支）带来的损害。

（九）使工程师满意的义务

使工程师满意（satisfaction of the engineer）或者使建筑师满意在FIDIC（1987版)、JCT98合同中都有类似的规定。但是，FIDIC“新红皮书”（1999版）取消了FIDIC合同1987年第四版的中“使工程师满意”规定，没有出现“使工程师满意”的措辞。

第二章 招 标 投 标 信 函

第一节 招 标 投 标 概 述

一、招标投标含义

在国际工程建设实践中，招标和投标是国际工程合同达成必经的两个阶段。招标（Invite Tenders）指的是对拟发包工程的内容标明，它主要包括项目概况、数量和质量要求，以及采用的图纸和规范、标准等。因此招标是指雇主（Employer）标明其拟发包工程的内容、要求等，以招引或邀请某些愿意承包并符合投标资格的承包者（Contractor）对承包该工程所采用的施工方案和要求的价格等进行投标，通过比价而达成交易的一种经济活动。投标（Submission of Tender）是与招标相对应的概念，它是指投标人应招标人的邀请或投标人满足招标人最低资质要求而主动申请，按照招标的要求和条件，在规定的时间内向招标人递价，争取中标的行为。实践中，中标的投标人成为承包商。

国际工程承包招标投标使得承包商与工程师开始接触，这个阶段根据工程性质的不同，可能在短时间内完成，也可能会持续很长时间。投标也是承包商和雇主接触的第一个阶段，只有中标的承包商才有可能就所投标的国际工程与雇主达成协议。这期间将发生大量承包商与工程师和雇主之间的信函往来，以期解决招标投标关涉的一些问题。

二、国际招标投标中注意的问题

近年来，随着海外工程承包业务规模的迅速拓展，国内公司在海外工程中所面临的市场环境的复杂性和风险的不可预见性大大增加，相应的法律风险也日益增加。因此，承包商必须了解海外工程适用的国际惯例和当地法律环境，总结经验教训，在签约前的投标阶段，加强投标风险的审查，以便及早识别并控制、规避与防范招标投标风险，防患于未然。

（一）项目资金来源

资金来源通常主要有两个方向：1）是否属世行、亚行或其他国际组织贷款项目？如果是世行、亚行或其他国际组织贷款项目，投标人可以大胆参加，通常此类项目工程款的回收是有保障的。2）是否由项目所在国政府出资？如果是项目所在国政府出资，则要充分考虑该政府的经济状况、偿还信誉和能力等方面，从而决定是否参加该项目。

（二）合约的性质和特殊条款

当拿到标书后，投标人要明确标书规定的合约性质：是单价合约，还是总价合约；是有条件的合约，还是条件待定合约。另外，在国际工程招标中，通常采用FIDIC条款作为合约的基本条款。雇主会根据实际情况，针对不同的项目，制定特殊条款，对此承包商应予以注意。

（三）中标文件与合同文件一致，投标时间充足

审查中标文件与合同文件一致性是避免开工后发生纠纷的重要工作。同时，充足的投标时间可以使承包商充分理解并评审招标文件和合同文件，避免不准确定义和合同条款不一致引起的纠纷。

（四）投标前应确保雇主对项目资金有合理的来源和安排。

有的项目投标人没有认真了解雇主的资信和项目资金状况，就盲目投标。合同签署后，才发现雇主其实并没有足够的项目资金，导致项目无法正常开工或者中间停工。因此，为避免出现此类情况，建议投标人在投标前应通过信函往来确保雇主对项目资金有合理的来源和安排。

（五）投标前应对当地建筑标准和规范进行了解

有的项目投标时只处于概念设计或方案设计阶段，设计深度不够，诸多设计参数都没有确定，承包商对项目所在国的建筑规范并不了解，无法编制详细的工程量清单，项目存在较大的潜在风险。因此，建议承包商在投标前与工程师和雇主进行信函沟通，要求提供足够深度的图纸，投标前对当地的建筑标准和规范进行必要的了解，特别是图纸设计深度不够的项目更要加强了解。

（六）注意价格与支付条款

合同价格与支付条款对于承包商来说是合同最重要的条款之一，承包商承包工程的最主要目的就是获得合理的合同价格，得到及时的付款；而雇主为了确保工程期限和质量，一般会通过合同价格和支付条件限制承包商。

（七）合同纠纷解决方式是否合理

纠纷解决条款在承包商评审合同时经常被忽略，该条款也是承包商最容易向雇主让步的条款之一，因为承包商一般觉得合同关于纠纷解决方式的规定只是摆设。但是，一旦发生纠纷，如果没有合理的纠纷解决方式，承包商会面临无法克服的法律障碍。

第二节 招标投标信函

Letter 1

To engineer, requesting inclusion in list of tenderers

Dear

We were interested to note from [*state source*] that tenders are to be invited for the above project in the near future.

This is a type of work in which we are very experienced and we should welcome an invitation to tender. The following information may be of assistance to you if you are not already aware of our capabilities:

[List the following information:]

1. *Names and addresses of all directors.*
2. *Address of registered office.*
3. *Website.*
4. *Share capital of firm.*

5. *Annual turnover during the last three years.*
6. *Number and positions of all office-based staff.*
7. *Number of site operatives permanently employed in each trade.*
8. *Number of trained supervisory staff.*
9. *Number and value of current contracts on site.*
10. *Address, date of completion and value of three recently completed projects of similar character to that for which tenders are to be invited.*
11. *Names and addresses of clients, engineers or quantity surveyors connected with the projects noted in 10 above and to whom reference may be made.*

We look forward to hearing from you in due course.

Yours faithfully

信函 1

致工程师，要求参与投标

敬启者：

我方从［注明信息来源］了解到贵方不久将邀请投标人参与上述项目投标。

我方对于这类项目非常有经验，希望收到贵方投标邀请函。如对我公司的实力尚不了解，可参阅以下信息。

［列出以下信息］

1. 所有董事的姓名和地址。
2. 公司注册地址。
3. 网址。
4. 公司股本。
5. 最近三年的年营业额。
6. 所有在册员工的人数和职务。
7. 每一专业工种永久雇用的现场施工人员名单。
8. 经过培训的监督管理人员数量。
9. 目前现场合同的数量及其价值。
10. 与受邀参与投标项目相似，且为近期完工的三个项目的地址，竣工日期与项目金额。
11. 上述第10条中项目相关的客户，工程师，成本估算师的姓名，地址以及联系人。

期待届时收到贵方回复。

敬上

Letter 2

To engineer, if no response to request for inclusion on list of tenderers

Dear

We refer to our letter of the [*insert date*], requesting inclusion on the list of tenderers for the above project.

Since we have not heard from you, we take this opportunity to re-affirm our interest in the project and assure you of our experience in work of this nature.

We should be delighted to meet you to expand upon the details given in our earlier letter. We have the facilities to make a presentation showing recent projects we have carried out which may be of interest to your client.

Our managing director [*or insert appropriate designation*], M. . . [*insert name*], will telephone you on [*insert day*] .

Yours faithfully

信函 2

致工程师，如果要求参与投标，但未获回复

敬启者：

我方于［插入日期］致函贵方申请参与上述项目投标。

由于我方至今未收到贵方答复，故借此机会再次确认我方对该项目感兴趣，并请贵方相信我方对此类工程施工有丰富的经验。

我方期待与贵方会晤，并就上次信中提及的细节进行深入讨论。我方有设施可以展示近期已经施工的工程项目，贵方客户可能会感兴趣。

我方总经理［或填入适当头衔］［填入姓名］将于［插入日期］与贵方电话联系。

敬上

Letter 3
To engineer, agreeing to tender

Dear

Thank you for your letter of the [*insert date*] from which we note that you intend to invite tenders for the above project.

We should be pleased to be included on the tender list. No doubt you will be sending further details in due course.

Yours faithfully

信函 3
致工程师，同意参与投标

敬启者：

感谢贵方［插入日期］来函，我方已知悉贵方将对上述项目邀标。

若我方入围投标，将深感荣幸。相信贵方届时会将更多细节发给我方。

敬上

Letter 4
To engineer, if contractor unwilling to tender

Dear

Thank you for your letter of the [*insert date*] from which we note that you intend to invite tenders for the above project.

With regret, we must ask to be excused from tendering on this occasion due to our very heavy workload. We do hope, however, that you will give us the opportunity to tender for other projects on other occasions in the future.

Yours faithfully

信函 4
致工程师，如果承包商不愿投标

敬启者：

感谢贵方［插入日期］来函，我方已知悉贵方将对上述项目邀标。

由于目前工程任务繁重，无法参与贵方招标，我方深感遗憾。但是，我方真切希望将来有机会参与贵方其他工程项目的投标。

敬上

Letter 5

To engineer, if contractor asked to provide information prior to inclusion on tender list

Dear

Thank you for your letter of the [*insert date*] setting out brief details of the above project and requesting particulars of this company.

On the information provided, it appears that the project would be directly related to our skills and experience and we should be delighted to be included on the tender list. The information you require is as follows:

[List answers using same numeration as in engineer's letter.]

Yours faithfully

信函 5

致工程师，在入围投标人名单之前，如果要求承包商提供信息

敬启者：

感谢贵方［插入日期］来函说明上述项目的简要细节并要求提供公司详细情况。

根据贵方提供的信息，该项目与我方的施工技术与经验直接相关。如能入围投标，我方将深感荣幸，贵方要求的有关信息如下：

［根据工程师信函中提出问题的序号逐项给予答复。］

敬上

Letter 6

To engineer，if the contractor is informed that the tender date is delayed and is still willing to submit tender

Dear

Thank you for the letter of the [*insert date*] informing us that the date for despatch of tender documents has been revised to [*insert date*]. We confirm that we are still willing to submit a tender for this project.

Yours faithfully

信函 6

致工程师，如果承包商获悉投标延期，仍愿意递标

敬启者：

感谢贵方［插入日期］来函通知我方投标截止日期改为［插入日期］。我方确认仍愿意投标该项目。

敬上

Letter 7

To engineer, if the contractor is informed that the tender date is delayed and is unwilling to tender

Dear

Thank you for your letter of the [*insert date*] informing us that the date of despatch of tender documents has been revised to [*insert date*]. We regret that it will be impossible for us to rearrange our very heavy workload so as to be able to submit a tender in accordance with the new timetable.

We do hope, however, that you will give us the opportunity to tender for other projects on other occasions in the future.

Yours faithfully

信函 7

致工程师，如果承包商得知投标延期，不愿参加投标

敬启者：

感谢贵方［插入日期］来函通知我方投标文件报送截止日期已更改至［插入日期］。我方无法重新安排繁重的工程任务，不能按照贵方新的时间安排递标，对此我方深表遗憾。

但是我方真切希望将来有机会参与贵方其他工程项目的投标。

敬上

Letter 8

To engineer, acknowledging receipt of tender documents

Dear

Thank you for your formal invitation to tender for the above project with which you enclosed [*list documents enclosed*].

We confirm that we will submit our tender by the [*insert tender date*].

[*If appropriate, add:*]

We wish to inspect the detailed drawings and visit site. Our M... [*insert name*] will telephone you to make the necessary appointment within the next few days.

Yours faithfully

信函 8

致工程师，投标人确认收到投标文件

敬启者：

非常感谢贵方正式邀请我方参加上述工程项目的投标，我方已收到随函文件［列出附函文件的名称］。

我方确认会在［插入投标日期］前提交投标书。

［如适合，请补充：］

我方希望查看详细图纸并且考察施工现场。我方［插入姓名］先生（或女士）将在最近几日与贵方电话预约。

敬上

Letter 9

To engineer，regarding questions during the tender period

Dear

We have carefully examined the tender documents enclosed with your letter of the [*insert date*]. We have examined the detailed drawings at your office and we have visited site. There are certain items which require clarification as follows：

[*List items requiring clarification.*]

Items marked with a red × are urgent and，if we are to meet the date for submission of tenders，we need clarification of these points by [*insert date*].

Yours faithfully

信函 9

致工程师，关于投标期间的问题

敬启者：

我方仔细阅读了［插入日期］信函中所附招标文件。我方已经在贵方办公室查阅了工程详图，并考察了施工现场。我方有以下问题需要贵方澄清：

［列出所需澄清的事项。］

标有红色×符号的为紧急事项。如果我方必须如期提交标书，我方需要贵方在［插入日期］前澄清这些事项。

敬上

Letter 10

To engineer, requesting extension of tender period

Dear

We are preparing our tender for the above project with the greatest possible speed. Prices for a number of the sub-contract items, however, will not be in our hands until after the date for submission of tenders. Clearly, we will be unable to submit a tender unless the tendering period is extended. We therefore request an extension of the period by [*insert period of extension, which should be as short as possible*]. We are proceeding on the assumption that you will be able to grant our request, but if you feel unable so to do, please let us know immediately so that we can stop what will become abortive work.

Yours faithfully

信函 10

致工程师，要求延长投标期限

敬启者：

我方正以最快速度准备上述项目的投标书。然而，我方只有在递标日期以后才能拿到分包合同报价。显然，如果不延长投标期限，我方则无法按期提交标书。因此，我方要求将投标期限延长［插入投标延长期限，越短越好］。如果贵方同意我方请求，我方将继续投标。但是，如果贵方无法延长投标期限，请及时通知我方，以便我方停止无效工作。

敬上

Letter 11

To engineer，withdrawing qualification to tender

Dear

In response to your letter of the [*insert date*], we confirm that we withdraw the qualification to our tender dated [*insert date*] without amendment to the tender sum of [*insert amount*].

The qualification to which we refer above is:

[*Set out the qualification using the precise wording used in the tender.*]

Yours faithfully

信函 11

致工程师，撤回投标资格

敬启者：

答复贵方［插入日期］来函，我方确认撤回［插入日期］的投标资格，不修改［插入金额］的投标金额。

我方上述提到的资格如下：

［请使用标书中的准确表达方式列出所撤销的资格。］

敬上

Letter 12

To engineer, if confirming offer where the overall price is dominant

Dear

Thank you for your letter of the [*insert date*] with which you enclosed a list of errors detected in our pricing of the bills for the above project.

We have carefully examined the list and we note that, in accordance with the CIB Code of Practice for the Selection of Main Contractors, the overall price is to be dominant in this instance. Therefore, please take this as notice that we confirm our offer of [*insert amount*] as stated in our tender dated [*insert date*].

We note what you state regarding endorsement and we agree to its terms.

Yours faithfully

信件 12

致工程师，如果确认以总价为主的报价

敬启者：

感谢贵方［插入日期］来函，函中列出我方上述工程项目报价中的错误。

我方仔细检查了贵方所列错误清单，我方注意到依据《CIB 总承包商选定法规》，此情况应以总报价为主。因此，通知贵方我方确认［插入日期］的标书中的报价［插入报价金额］为我方报价。

我方注意到并同意贵方提出的相关说明。

敬上

Letter 13

To engineer, if withdrawing offer where the overall price is dominant

Dear

Thank you for your letter of the [*insert date*] with which you enclosed a list of errors detected in our pricing of the bills for the above project.

We have carefully examined the list and we note that, in accordance with the CIB Code of Practice for the Selection of Main Contractors, the overall price is to be dominant in this instance. In view of the nature of the errors, we regret that we must withdraw our offer.

Yours faithfully

信函 13

致工程师，如果撤回以总价为主的报价

敬启者：

感谢贵方［插入日期］来函，函中列出我方上述工程项目报价中的错误。

我方仔细检查了贵方所列错误清单，我方注意到依据《CIB 总承包商选定法规》，此情况应以总价为主。鉴于错误的性质，我方必须撤回报价，对此深感遗憾。

敬上

Letter 14

To engineer, if amending offer where the pricing document is dominant

Dear

Thank you for your letter of the [*insert date*] with which you enclosed a list of errors detected in our pricing of the bills for the above project.

We have carefully examined the list and we note that, in accordance with the CIB Code of Practice for the Selection of Main Contractors, the pricing document is dominant in this instance. In view of the nature of the errors, we have amended our offer. Our amended tender price is [*insert amount*] and we enclose details of the relevant calculations.

Yours faithfully

信函 14

致工程师，如果在以定价文件为主的情况下修改报价

敬启者：

感谢贵方［插入日期］来函，函中列出我方上述工程项目报价中的错误。

我方仔细检查了贵方所列错误清单，我方注意到依据《CIB 总承包商选定法规》，报价应以定价文件为主。鉴于错误的性质，我方已经修改了报价，将其修改为［插入金额］。此外，我方附上相关算法的详情。

敬上

Letter 15

To engineer, if tender accepted (a)

Dear

Thank you for your letter of the [*insert date*] accepting our tender of the [*insert date*] in the sum of [*insert amount*] for the above Works in accordance with the drawings numbered [*insert numbers*] and the bills of quantities [*or specification/work schedules*].

We understand that a contract now exists between the employer and ourselves and we look forward to receiving the contract documents for signing/execution as a deed [*delete as appropriate*] in due course.

Yours faithfully

信函 15

致工程师，如果中标（a）

敬启者：

感谢贵方于［插入日期］来函告知接受我方对上述项目［插入日期］，报价为［插入金额］的投标，该投标的依据是编号为［插入号码］的图纸与工程量清单［工程说明/工程进度表］。

我方与雇主已形成合同关系，因此我方期待届时收到合同文件以便签署/立契约［酌情删除］。

敬上

Letter 16

To engineer, if tender accepted (b)

Dear

Thank you for your letter of the [*insert date*] accepting our tender of the [*insert date*] in the sum of [*insert amount*] for the completion of the design and the construction of the above project in accordance with the Employer's Requirements, the Contractor's Proposals and the Contract Sum Analysis.

We understand that a contract now exists between the employer and ourselves and we look forward to receiving the contract documents for signing/execution as a deed [*delete as appropriate*] in due course.

Yours faithfully

信函 16

致工程师，如果中标（b）

敬启者：

感谢贵方［插入日期］来函告知接受我方［插入日期］，报价为［插入金额］的投标，我方依照雇主要求，完成承包商建议书以及合同总价分解方案，设计并完成该项目施工。

我方与雇主已形成合同关系，因此我方期待届时收到合同文件以便签署/立契约［酌情删除］。

敬上

Letter 17

To engineer, if tender accepted (c)

Dear

Thank you for your letter of the [*insert date*] accepting our tender of the [*insert date*] in the sum of [*insert amount*] for the construction of the above project and the design and construction of the Contractor's Designed Portion in accordance with the drawings numbered [*insert numbers*] and the bill of quantities [*or specification/work schedules*] and the relevant Employer's Requirements, Contractor's Proposals.

We understand that a contract now exists between the employer and ourselves and we look forward to receiving the contract documents for signing/execution as a deed [*delete as appropriate*] in due course.

Yours faithfully

信函 17

致工程师，如果中标（c）

敬启者：

感谢贵方［插入日期］来函告知接受我方［插入日期］，报价为［插入金额］的投标，我方按照［插入编号］图纸，工程量清单［工程说明/工程进度表］，雇主要求，相关承包商建议书，完成上述项目的施工与设计，以及承包商设计部分的施工。

我方与雇主已形成合同关系，因此我方期待届时收到合同文件以便签署/立契约［酌情删除］。

敬上

Letter 18
To engineer，if purporting to accept tender

Dear

Thank you for your letter of the［*insert date*］, which purports to be an acceptance of our tender of the［*insert date*］for the Works.

It is clear that your letter is not an acceptance of our tender，indeed it attempts to insert the following changes：［*list the changes*］.

Therefore，your letter is，strictly，a counter-offer and no contract exists unless we accept it. In the circumstances，we have no intention of accepting it. We are advised that your counter-offer had the effect of rejecting our original tender. Therefore，we confirm that we will re-open our tender for a further［*insert period*］to allow you to unequivocally accept it.

Yours faithfully

信函 18
致工程师，如果有意向接受投标

敬启者：

感谢贵方［插入日期］来函告知贵方有意接受我方于［插入日期］发出的项目投标。

很明显贵方的来函并非我方投标的接受函，该函试图进行以下变更［列出变更事项］。

因此，严格意义上讲，贵方来函旨在还价，如果我方不接受，就不存在合同关系。在此情形下，我方不打算接受贵方的还价。我方认为贵方的还价等同于贵方拒绝了我方原来的投标方案。因此，我方确认将会在未来［插入时间段］重新投标，以便贵方明确接受。

敬上

Letter 19

To engineer，if another tender accepted

Dear

Thank you for your letter of the［*insert date*］from which we note that our tender was not successful in this instance.

We await details of the full list of tender prices with interest and assure you of our willingness to receive your future enquiries.

Yours faithfully

信函 19

致工程师，接受其他投标

敬启者：

感谢贵方［插入日期］来函，我方获悉此次未中标。

我方期待收到完整的投标价格详单，并保证我方愿意接受贵方日后询价。

敬上

第三章　合同签订及履行前的准备信函

第一节　合同签订及履行前的准备概述

一、国际工程合同文件

（一）国际工程合同文件含义

国际工程合同是竞争性招标程序中雇主和承包商谈判的结果，而国际工程合同文件则是对招标投标结果和合同条件的书面表达。合同文件需要合同双方即雇主和承包商的签字认可方能生效，书面签字协议已经成为各国法律普遍接受的实践做法。达成建设工程合同的基本目的是准确界定和明确合同双方的权利和义务。国际工程的复杂性决定了国际工程合同一般会采用繁复冗长的形式以应对精确描述法律、金融、技术的各项规定。

如果承包商在投标阶段获得成功，雇主或工程师将以信函的方式通知这一事实。每个参与投标的建筑公司都希望从招标人那里得到中标通知书，而现实的情况是，在众多投标者中，只有一个投标人能获此殊荣，与招标人——雇主签订合同。

包含基本信息的中标函是签订正式工程承包合同的"入门证"，领了入门证，方能入门。但是合同签订过程中仍有很多细节问题需要仔细斟酌，因此，承包商不应轻易签字来完成合同文件，在签字前，应对合同文件和信函作认真的审查。对于印制的合同，一定要检查其条款是否与招标文件内容一致或者印制合同与招标文件差异的解释。合约图纸也应与投标图纸相一致。在传统的采购条件下，从招标开始到合同文件签署，工程师会对图纸及规范进行修改。如果承包商发现了合同文件，包括图纸、规范的不一致，一定要积极进行信函沟通，直至所有的文件条款、图纸、规范纸面上一致之后再签字，任何口头承诺都不能成为签字的理由。

（二）雇主与承包商对合同文件理解的一致性

雇主的要求与承包商建议的一致性是至关重要的。一旦合同签字，用合同方法或司法手段来解决合同条款的不一致性都是低效率的或者增加费用的。同时必须注意的是，在正式合同文件签字前投标时的图纸已被招标人的工程师进行了修改，而在合同文件正式签字后，对修改部分的责任需由承包商来承担，而不是雇主聘用的工程师。

所有的合同文件都是为了主合同的解释目的而存在的，这些合同文件之间相互支撑，构成完整的合同体系。一般情况下，缔约各方的合同目的并不是在签订合同之初的谈判和协商中实现的，而是通过最终合同履行来实现的。对合同的可行解释应根据法律规定进行。根据法律推定原则，合同签约各方都应当知道他们想要什么，在合同中表达了他们想表达的意思和观点，且合同当事人也理解他们所说的，并为此承担后果。因此，合同各方有义务和责任在履行合同前阅读和理解合同内容，否则就没有理由和借口违背合同的约定。

当合同条款发生冲突时，特殊条款的效力优于普通条款，手写条款优于印刷条款；如果合同中文字和数字表达不一致，文字表达效力优先；如果有含混不清、模棱两可的条款，法院会作出不利于合同提供方的解释；如果有图纸和规范之间存在冲突，应该注意的是，法院一再强调以规范（Specification）为准。因此，在处理合同文件不一致性问题上，使用正确的商业信函是非常重要的。

（三）国际工程合同文件组成

国际工程合同由许多不同的文件组成，下面的一系列文件是招标、谈判和施工过程中必不可少的，这些文件根据实践的不同组合，构成国际工程合同的主要内容：（1）投标邀请函；（2）投标人须知；（3）一般条款；（4）附加条款；（5）技术条件；（6）图纸；（7）附录；（8）提案；（9）投标押金；（10）协议；（11）履约保证金；（12）工料支付保证金。

二、意向书

（一）意向书的含义与特征

招标是先合同阶段的程序，在这个阶段，合同各方需要彼此熟悉，其中一个简单的方法就是向对方提出问题，或者希望对方澄清一些与文件和沟通相关的问题。这就要求合作各方进行对话与讨论，创造一种公平竞争的氛围。在大多数情况下，雇主会组织所有承包商进行面对面的会议来回答和解疑问题，通过接触相互了解。在一些情况下，投标说明可能会在雇主与中标承包商间进行，也可能在雇主与所有承包商间进行。在做出最后选择之前，各方会通过意向书进行沟通，逐步锁定最终协议的主要内容。在实践中，错误理解意向书的法律价值时有发生，因为在所有建筑施工标准合同中，包括 FIDIC、LOGIC and the Dutch CMM Contract 等，都没有意向书的相关条文。一般来说，意向书是承载当事人各方认同的条件或条款的文件，如果任何一方想在意向书的基础上达成最终协议，应克服有争议的问题，尽管最终消除所有争议，协议也不一定能够达成。

意向书（A Letter of Intent，LOI）是商务活动中的双方或多方在进行合作之前，通过初步谈判，就合作事宜表明基本态度、提出初步设想的文书。意向书主要是表达合作各方共同的目的和责任，是签订合同前达成意向性、原则性一致意见的表示。它是实现实质性合作的基础。意向书既可以使磋商合作的步伐走得稳健而有节奏，避免草率从事，盲目签约，也可以及时抓住意向、开拓发展，避免失去商机。

（二）意向书的法律效力

意向书是一方当事人向另一方当事人发出的表达初步合作设想的信函，如果意向书中列明，对方签字即达成合同，发出意向书的一方在一定时间内将在法律上受到约束，不能再与其他方签订合同。因此，意向书的内容决定了意向书是否具有法律约束力。实践中，在确定意向书是否对双方具有约束力时，一要看意向书的具体内容，二要看接到意向书的一方对意向书内容所作的反应。如果发出意向书的一方在意向书中有意限制了其法律效力，则通过意向书的商谈和合作并不具有法律效力。但是，如果意向书中包含了关键问题的达成与许诺，例如保密条款、诚信条款、a stand-still or no-shop provision 条款（这个条款的含义是在卖方和潜在买方之间达成的禁止卖方在一定时间内将货物或服务卖给其他买方的约束性条款）一旦达成，意向书各方将受到法律的约束。如果意向书的内容中没有约束性条款，意向书则不具有法律约束力。

（三）意向书在建筑工程承包项目中的作用

对于一个建筑工程，雇主与承包商、承包商与分包商最终应将工程建设的各个方面及履行细节确定下来。这些工作的完成是需要大量的合同和协商工作的，并且将耗费大量的人力和物力。在这些工作完成前，承包商不会动工，雇主也不会支付工程费用。但是在现实中，项目各方迫于商业压力在签订正式合同前都会着手准备并开始工作，借助于意向书的某些约定开始先期工作。意向书可以推动建筑项目先行一步，例如开始设计或寻找合适的分包商等。但是如前所述，在没有达成正式工程承包合同前，通过意向书部分条件的有效约定开始进入工作状态的行为，或多或少都存在一定的风险。

三、承包商履约保函或保证金

履约保函（Performance Guarantee）是指应承包商（申请人）的请求，银行金融机构向工程的投资人——雇主作出的一种履约保证承诺。如果承包商日后未能按时、按质、按量完成其所承建的工程，则银行将向雇主方支付一笔约占合约金额5%～10%的款项。履约保函有一定的格式限制，也有一定的条件。

履约保函将在很大程度上促使承包商履行合同约定，完成工程建设任务，从而有利于保护雇主的合法权益。一旦承包商违约，担保人要代为履约或者赔偿经济损失。履约保证金额的大小取决于招标项目的类型与规模，但必须保证承包人违约时，雇主不受损失。在投标须知中，发包人要规定履约担保的形式，中标人应当按照招标文件中的规定提交履约担保。

根据《中华人民共和国招标投标法实施条例》（中华人民共和国国务院令第613号）第五十八条，招标文件要求中标人提交履约保证金的，中标人应当按照招标文件的要求提交。履约保证金不得超过中标合同金额的10%。

与履约保函比较，履约保证金是以现金形式提供的担保，两者都是雇主方为保障工程建设的顺利履行要求承包商提供的。履约保函或履约保证金的提供在某种意义上证明了承包商的实力和违约救济的保障。两者相比，虽然履约保函的办理手续复杂，但是对于承包商而言，提供保函会更有利一些。

四、承包商人员与项目经理

项目经理是工程项目施工组织的总负责人，是国际承包工程公司在国外的代理人。项目经理在某种意义上也是项目管理团队的统称，这个团队包括管理人员、技术人员、行政人员，以及财务、物资、合同和工程技术方面的专家。其工作对象既涉及外国技术人员和工人，也涉及外国的法律、海关、财政和税务等问题。因此，作为一名国际承包工程的项目经理，要具备一些基本的素质，才能胜任工作。FIDIC“新红皮书”第4.3款规定，承包商应任命承包商代表，并授予他在按照合同代表承包商工作时所必需的一切权力。除非合同中已注明承包商代表的姓名，否则承包商应在开工日期前将其准备任命的代表姓名及详细情况提交工程师，以取得同意。如果同意被扣压或随后撤销，或该指定人员无法担任承包商的代表，则承包商应同样地提交另一合适人选的姓名及详细情况以获批准。

因此，承包商在确定其现场管理代表后，应及时与雇主或工程师沟通，以方便工作的衔接和开展。但是承包商代表一旦指定，承包商就无权任意撤换，除非经过雇主或工程师的同意。

第二节　合同签订及履行前的准备信函

Letter 1
To engineer, returning contract documents
Special delivery

Dear

Thank you for your letter of the [*insert date*] with which you enclosed the contract documents. We have pleasure in returning them herewith, duly signed/executed as a deed [*delete as appropriate*] as requested.

We look forward to receiving a certified copy of the contract documents within the next few days.

Yours faithfully

信函 1
致工程师，返还合同文件
快递

敬启者：

感谢贵方［插入时间］来函以及随函所附的合同文件。今特此返还合同文件，并按要求妥为签字/立契约［酌情删除］。

我方期待在未来几天内收到合同文件的核证副本。

敬上

Letter 2

To engineer, if mistakes in contract documents and previous acceptance of tender

Dear

We are in receipt of your letter of the [*insert date*] with which you enclosed the contract documents for us to sign/execute as a deed [*delete as appropriate*].

There is an error on [*describe nature of error and page number of document*]. This is not consistent with the tender documents on which our tender is based which, together with your acceptance of the [*insert date*], forms a binding contract between the employer and ourselves.

We therefore return the documents herewith and we look forward to receiving the corrected documents as soon as possible.

Yours faithfully

信函 2

致工程师，如果合同文件和以前的中标函有错误

敬启者：

我方收到贵方［插入日期］来函以及随函所附要求我方签署/立契约的合同文件［酌情删除］。

在［说明错误性质以及文件的页码］出现一处错误。这与我方投标时所依据的投标文件不符，该文件与贵方［插入日期］接受投标函共同构成雇主与我方之间具有约束力的合同关系。

所以，我方特此返还合同文件，并且期望尽快收到更正后的合同文件。

敬上

Letter 3

To engineer, if mistakes in contract documents and no previous acceptance of tender

Dear

We are in receipt of your letter of the [*insert date*] with which you enclosed the contract documents for us to sign/execute as a deed [*delete as appropriate*].

There is an error on [*describe nature of error and page number of document*]. This is not consistent with the tender documents on which our tender is based and we are not prepared to enter into a contract on the basis of the contract documents in their present form.

We therefore return the documents herewith and we look forward to receiving the corrected documents as soon as possible.

Yours faithfully

信函 3

致工程师，如果合同文件中有错误，且之前未接受投标

敬启者：

我方已收到贵方［插入日期］来函以及随函所附要求我方签署/立契约的合同文件［酌情删除］。

在［说明错误性质以及文件的页码］出现一处错误。这与我方投标时所依据的投标文件不符，我方不准备根据目前合同文件形式签订合同。

因此，我方特此返还合同文件，并且期望尽快收到更正后的合同文件。

敬上

Letter 4

To engineer, if contractor asked to commence before contract documents signed, but tender accepted

Dear

Thank you for your letter of the [*insert date*] from which we note that the employer requests us to commence work on site pending completion of the contract documents.

It is our understanding of the situation that we are already in a binding contract with the employer on the basis of our tender of the [*insert date*] and an acceptance of the [*insert date*] on terms incorporated by such tender and letter of acceptance.

If the employer will send us written confirmation indicating agreement with our understanding of the situation as expressed in this letter, we will be happy to commence as requested.

Yours faithfully

信函 4

致工程师，如果要求承包商在签合同之前开工，但已接受投标

敬启者：

感谢贵方［插入日期］来函，获悉雇主要求我方在签订合同前开始现场施工。

在此情况下，依据我方［插入日期］的投标书与［插入日期］的中标函，就该标书与该中标函所包含的条款，我方认为已与雇主之间形成有约束力的合同关系。

如果雇主向我方发出书面确认函，表明同意我方对本函所述情况的理解，我方将按要求开工。

敬上

Letter 5

To engineer, if contractor asked to commence before contract documents signed and tender not accepted

Dear

Thank you for your letter of the [*insert date*] from which we note that the employer requests us to commence work on site pending completion of the contract documents.

We note that the employer has not yet accepted our tender of the [*insert date*] and, therefore, there is currently no contract between us, no mechanism for payment, instructions etc. We are sure that you will understand our reluctance to commence work under these circumstances.

Obviously, if the employer will send us written and unequivocal acceptance of our tender or executes the contract documents, we will be happy to commence as requested.

Yours faithfully

信函 5

致工程师，如果要求承包商在签署合同之前开工，但尚未接受投标

敬启者：

感谢贵方［插入日期］来函，获悉雇主要求我方在签订合同前开始现场施工。

我方获悉雇主并未接受我方［插入日期］的投标，因此，目前双方并不存在合同关系，也不存在支付，指示等事宜。在此情况下，我方不愿开工，相信贵方能理解。

当然，如果雇主以书面形式明确告知我方中标，或执行合同文件，我方愿意按要求开工。

敬上

Letter 6

To engineer, if contract not signed and certification due

Special delivery

Dear

We tendered for the above work on the [*insert date*] in the sum of [*insert amount*] and the employer accepted our tender unequivocally by letter dated the [*insert date*]. A binding contract exists between the employer and ourselves in terms incorporating, among other things, the provisions of [*insert the full title of the form of contract*] which provides for certification of monies due to us.

Although we have difficulty in understanding why you have not yet prepared the formal contract documents for signature/completion as a deed [*delete as appropriate*], the completion of such formal contract documents will simply reflect the respective rights and obligations of the parties as already agreed and about which there is no doubt.

We, therefore, require you to issue your certificate in accordance with clause ____ of the conditions of contract

If such certificate is not in our hands by [*insert date*] we will take immediate legal action against the employer for the breach.

Yours faithfully

信函 6

致工程师，如果未签署合同，且应该发出证明

快递

敬启者：

我方于［插入日期］以［插入金额］投标上述项目。在［插入日期］雇主来函明确告知接受我方投标。

依据相关合同条款，雇主与我方之间存在有约束力的合同关系，除其他事项外，其中［插入合同形式的完整标题］有条款规定应给我方提供资金证明。

虽然我方很难理解贵方为何还没有准备好正式合同文件，以便签署/立契约［酌情删除］，但是，签署正式合同文件将会体现双方之前已经达成一致的各自权利及义务，且对此毫无疑问。

因此，我方要求贵方根据合同____条款颁发证明。

如果［插入日期］前我方没能收到该证明，我方将立即采取法律措施起诉雇主违约。

敬上

Letter 7

To engineer, on receipt of letter of intent

Dear

We are in receipt of a letter of intent dated [*insert date*].

[*If not prepared to proceed, add*:]

The tender documents gave no indication that you might ask us to proceed on the basis of a tender of intent and we are not prepared to do so. It is far too risky for all parties. We will proceed on the basis of an unequivocal acceptance of our tender or on the execution of the formal contract documents.

[*If prepared to proceed, add*:]

Working on the basis of a letter of intent is risky for both parties. We are prepared to commence work on the basis of your letter, but note that we expect formal contract documents to be prepared ready for execution by both parties within the next [*insert period*]. Failure to do so will cause us to reassess the situation at that stage and probably to cease all further work.

Yours faithfully

信函 7

致工程师，收到意向书

敬启者：

我方收到［插入日期］意向书。

［若不准备继续，请补充：］

招标文件并未表明贵方可以在意向投标基础上要求我方施工，我方不准备这样执行，这对各方都有风险。我方只有在明确中标或执行正式合同情况下才会开始施工。

［若准备继续进行，请添加：］

依据意向书开工对双方都有风险。我方准备依照中标函开工，但提醒贵方注意我方希望贵方在下个［插入时期］内提供双方准备执行的正式合同文件。如果贵方不提供，我方将重新评估当前情况，并可能停止所有进一步的工作。

敬上

Letter 8

To engineer, if contractor asked to sign a warranty

Dear

Thank you for your letter of the [*insert date*] with which you enclosed a form of warranty for signature.

[*Add either*:]

We note that the form is identical to that attached to the contract documents and we have pleasure in returning it duly completed as requested.

[*Or*:]

Although the contract documents call on us to complete a form of warranty, no particular form was specified and we regret that your suggested form is not acceptable.

[*Or*:]

We note that there was no requirement in the contract documents for us to complete a warranty. However, we enclose a suggested warranty which we would be prepared to complete if you will complete a warranty to us as the form attached.

Yours faithfully

信函 8

致工程师，如果要求承包商签署担保书

敬启者：

收悉［插入日期］来函，函中附寄了要求我方签署的担保书。

［或补充：］

注意到该担保书与合同文件所附格式相同，我方将按要求签署，及时返还贵方。

［或者：］

虽然合同文件要求我方签署担保书，但未明确具体格式。对此，我方不能接受贵方提出的担保书格式，深表遗憾。

［或者：］

我方注意到合同文件并未要求我方签署担保书，但是，若贵方愿意依据我方所附格式签署我方附寄的担保书，我方则也会签署该担保书。

敬上

Letter 9

To engineer, on receipt of defective third party rights notice

Dear

Thank you for your letter dated [*insert date*] which purports to be a notice under clause ____.

[*Add either*:]

Please note that the purchaser/tenant/funder [*delete as appropriate*] named in your notice cannot acquire third party rights under this contract because such purchaser/tenant/funder was not identified by name, class or description in part ____ of the contract particulars.

[*Or*:]

Please note that your notice is defective in that [*set out why it is defective*]. Consequently, the purchaser/tenant/funder [*delete as appropriate*] named in your notice cannot acquire third party rights under this contract.

Yours faithfully

信函 9

致工程师，收到有瑕疵的第三方权利通知

敬启者：

感谢贵方在［插入日期］依据条款____发给我方通知函。

［或者补充：］

请贵方注意，根据合同，通知中指定的买方/承租方/投资方［酌情删除］不能获得第三方权利，因为合同细节第____部分并未明确买方/承租方/投资方的姓名，类别和说明等信息。

［或者］

请贵方注意贵方的通知有瑕疵，原因是［列出原因］。因此根据合同，贵方通知中指定的买方/承租方/投资方［酌情删除］不能获得第三方权利。

敬上

Letter 10

To engineer, on receipt of request for warranty

Dear

Thank you for your letter dated [*insert date*] issued as a notice under clause ____.

[*Add either*:]

However, we should point out that we are not required to provide such a warranty/sub-contractor warranty [*delete as appropriate*] because the relevant details were not inserted in part ____ of the contract particulars.

[*Or*:]

We are putting the necessary measures in place to provide the warranty as requested.

Yours faithfully

信函 10

致工程师，收到担保要求

敬启者：

感谢贵方在［插入日期］依据条款____发给我方通知函。

［可以补充：］

但是，我方必须指出我方不需要提供此类担保/分包商担保［酌情删除］，因为合同第____部分并未明确相关细节。

［或者：］

我方正在采取必要措施，根据要求提供担保。

敬上

Letter 11

To engineer, if contractor asked to supply a performance bond

Dear

Thank you for your letter of the [*insert date*] requesting us to provide a performance bond.

[*Add either*:]

We note that a form of bond was included in the contract documents and we are arranging for a bond to be executed accordingly.

[*Or*:]

Although the contract documents call on us to provide a performance bond, no particular form was specified. The form you include with your letter is not acceptable and we are arranging to have the bond executed in a form which we find acceptable.

[*Or*:]

The contract documents do not require us to provide a performance bond and we decline to do so.

Yours faithfully

信函 11

致工程师，如果要求承包商提供履约保证金

敬启者：

收到贵方［插入日期］来函，要求我方提供履约保证金。

［请添加：］

我方注意到合同文件包括保证金的形式，我方正在据此安排执行。

［或者：］

虽然合同文件要求我方提供履约保证金，但未规定具体形式。我方不接受贵方随函所附的保证金支付形式，我方正以我方可接受的形式安排支付此保证金。

［或者：］

合同文件并未要求我方提供履约保证金。因此，我方拒绝提供。

敬上

Letter 12
To employer，if asked to execute a novation agreement

Dear

Thank you for your letter dated [*insert date*] with which you enclosed a novation agreement which you are asking us to execute. By this agreement，we would effectively take over the contract you have with the engineer [*or substitute the relevant consultant*] just as though he/she had been in contract with us from the beginning.

[*Add either*：]

This is the same document that we undertook to execute when we tendered for this project；therefore，we have had the relevant signatures appended and we return it herewith. Although we have kept a copy，we should be glad to receive the final version properly executed by all parties.

Yours faithfully

信函 12
致雇主，如果要求执行更替型协议

敬启者：

收到贵方［插入日期］来函，随函付寄要求我方执行更替协议。依照此协议，我方将有效接管贵方与工程师［或替换为有关顾问］所签订的合同，就如他/她起初就已经与我方签订合同一样。

［或者：］

此文件为我方投标此项目时承诺执行的同一文件，因此，我方已经附上签名，现在特此返还。我方虽留有一份该文件，但仍希望收到各方签字的最终版本。

敬上

Letter 13

To engineer, enclosing the construction phase plan

Dear

We have completed the construction phase plan so far as we are able on the basis of the information currently to hand. The purpose of the plan is to show the way in which the construction phase is to be managed and the important health and safety issues in this project.

We take the opportunity to enclose a full copy of the plan for your comments and use. Please note that it is not to be treated as a mere paper exercise. Rather, it is an important tool in the construction process and something which is a mandatory requirement under the Construction (Design and Management) Regulations 2007. You will see that the plan is divided into the following main parts:

- description of the project,
- management of the work,
- arrangements for controlling significant site risks,
- health and safety file.

Yours faithfully

信函 13

致工程师，附上施工阶段计划

敬启者：

依据目前我方所能掌握的信息，我方已制定完成施工阶段计划。该计划旨在说明如何管理该项目的施工阶段以及卫生和安全问题。

我方藉此附上一份该计划的完整版本，请贵方提出意见并使用。提请贵方注意切勿轻视该计划，该计划是施工过程中的重要工具，也是《2007 建筑法规》（设计与管理）中的一项强制性要求。该计划包括以下几个部分：

- 项目描述；
- 工程管理；
- 现场风险管控；
- 卫生安全文档。

敬上

Letter 14

To engineer, regarding person-in-charge or agent

Dear

This is to inform you that the person-in-charge of this project on site will be [*insert name*]. He/She [*delete as appropriate*] is competent and experienced in this type of work.

Please report your presence on site to the person-in-charge/agent [*delete as appropriate*] immediately on arrival for health, safety and security reasons.

Yours faithfully

信函 14

致工程师，关于负责人或代理人

敬启者：

我方特此通知贵方该工程项目现场负责人为［插入姓名］，他/她［酌情删除］有能力与经验管理此类工程。

出于卫生、安全原因，如贵方到达施工现场，请立即告知现场负责人/代理人［酌情删除］。

敬上

Letter 15

To engineer，regarding the appointment of a site manager

Dear

We propose to appoint [*insert name*] as site manager for this project. He/She [*delete as appropriate*] is competent and experienced in this type of work and we should be pleased to receive your written consent to this appointment as required by supplemental provision paragraph ____ of the contract.

Please report your presence on site to the site manager immediately on arrival for health, safety and security reasons.

Yours faithfully

信函 15

致工程师，关于任命现场经理

敬启者：

我方建议任命［插入姓名］为该项目的现场经理。他/她［酌情删除］有能力且经验丰富，足以胜任此类工程的施工管理。按照合同补充条款____的规定，我方希望收到贵方书面同意此任命。

出于卫生、安全原因，贵方人员到达现场时，请立即向现场经理报告。

敬上

Letter 16

To engineer, regarding consent to removal or replacement of site manager

Dear

[*If removal*:]

We refer to our telephone conversation of the [*insert date*] and confirm that we intend to remove [*insert name*] as site manager for the reasons we discussed. We understand that you accept those reasons and we should be pleased to receive your written consent, as required by supplemental provision paragraph ____ of the contract, to the appointment of [*insert name*] as site manager from [*insert date*].

[*If replacement*:]

The present site manager, [*insert name*], will be leaving on [*insert date*] and we should be pleased to receive your written consent, as required by supplemental provision paragraph ____ of the contract, to the appointment of [*insert name*] as site manager from that date.

Yours faithfully

信函 16

致工程师，关于同意免职或更换现场经理

敬启者：

[如果免职:]

我方于［插入日期］与贵方通话，确认打算更换现场经理［插入姓名］并讨论了理由。贵方接受以上理由。因此，按照合同补充条款____的规定，我方希望收到贵方书面同意自［插入日期］起，任命［插入姓名］为现场经理。

[如果更换:]

目前的现场经理［插入姓名］将于［插入日期］离职。按照合同补充条款____的规定，我方希望贵方书面同意，自该日期起，任命［插入姓名］为现场经理。

敬上

Letter 17

To engineer, if required to furnish names and addresses of operatives

Dear

Thank you for your letter of the [*insert date*].

In accordance with clause ____ of the conditions of contract, we have pleasure in enclosing a full list of the names and addresses of persons who are concerned with the works and the capacities in which they are so concerned. Please inform us if you require further particulars. They are all persons whom we employ and for whom we are responsible.

There may well be other persons not in our employ who are concerned with the works for whom we do not have, nor do we have the authority to obtain, such information.

Yours faithfully

信函 17

致工程师，如果需要提供现场施工人员的姓名与地址

敬启者：

感谢贵方于［插入日期］的来信。

根据合同条款____的规定，我方已随函附上参与工程的相关人员的姓名，地址以及各自技能的完整清单。如需进一步的详细资料，请通知我方。以上工作人员均为我方雇用人员，我方对其负责。

参与该工程但不在我方雇用范围内的其他人员，我方没有，也无权获得上述相关信息。

敬上

Letter 18

To engineer，if passes are required

Dear

In accordance with clause ____ of the conditions of contract，we enclose a list of persons requiring passes to secure admission to the site. Please inform us what，if any，other details you require in respect of each person.

Yours faithfully

信函 18

致工程师，如果需要通行证

敬启者：

按照合同条款____要求，我方附上一份需要通行证的人员名单，以确保他们可以进入施工现场。如贵方还需每位员工的其他详细信息，请通知我方。

敬上

Letter 19

To employer, regarding the employer's representative

Dear

[*Either*:]

We note that M... [*insert name*] has been appointed as the employer's representative under clause ____ of the contract.

[*Or*:]

We note that M... [*insert name*] has been appointed as project manager. As there is no provision for such a role in the contract, we assume that he/she [*delete as appropriate*] has been appointed as the employer's representative under clause ____ of the contract.

[*Then*:]

We should be grateful if you would confirm, as required by clause ____, that M... [*insert name*] will exercise all the functions ascribed to you in the contract and specify what if any exceptions there are.

Yours faithfully

信函 19

致雇主，关于雇主代表

敬启者：

[其一：]

我方获悉，根据合同条款____要求，____先生［插入姓名］已被任命为雇主代表。

[或：]

我方获悉［插入姓名］已被任命为项目经理。由于合同中未规定项目经理一职，因此，根据合同条款____，我方认定其已被认命为雇主代表。

[然后：]

如果贵方能够根据合同条款____确认［插入姓名］将代行合同赋予贵方的所有职责，并明确特殊情况的处置措施，我方将不胜感激。

敬上

Letter 20

To employer，regarding the project manager

Dear

We write simply in order to clarify something. We note that you have appointed a project manager. The contract makes no provision for such a person，the key people in the contract being the employer，the contractor，the engineer and the quantity surveyor. It is obvious，if the project is to proceed smoothly，that everyone should be aware of his or her role and the powers and duties that go with it.

We shall，therefore，assume that the project manager is your own representative and we should be grateful if you would advise him/her [*delete as appropriate*] of the situation bearing in mind that only the engineer may issue instructions，certificates and extensions of time. If the project manager attends site meetings，he/she [*delete as appropriate*] should not become involved unless invited to do so by the engineer or by us. We trust that you will agree the importance of getting these points clear at the commencement of the project if serious confusion is to be avoided.

If you disagree with our view of the situation，we should be grateful to receive your own views as soon as possible.

Yours faithfully

Copy：Engineer

信函 20

致雇主，关于项目经理

敬启者：

我方来函旨在澄清以下事项。我方获悉贵方已任命一名项目经理。但合同中并无该人员的任何规定，合同关键方为雇主，承包商，工程师以及估算师。显然，为项目进展顺利，各方都应清楚其职位以及相应的权利与义务。

因此，我方认定该项目经理仅为贵方代表，希望贵方能够明确告知他/她［酌情删除］施工现场情况，切记只有工程师可以发布指令，签发证明以及延长工期。如果该项目经理参加施工现场会议，未经工程师或我方邀请，该代表不可参与讨论现场会议的相关内容。

为了避免出现严重混乱情况，我方相信贵方会同意在项目开工时明确这些要点是至关重要的。

若贵方反对我方对此事的意见，我方则希望尽快收到贵方意见。

敬上

抄送至：工程师

第四章 保 险 信 函

第一节 保 险 概 述

国际工程项目通常都具有投资高、规模大、工期长等特点，易受自然灾害、暴乱、战争等事件影响，从而引起项目参与方巨大的经济损失。通过购买相应的商业保险，项目参与方可以将项目中的一些不可控风险有效转移，降低对雇主或承包商因工程风险造成的损失和不利影响。

保险信函主要发生在雇主与承包商之间关于投保责任分担的协商，雇主或承包商与保险人直接关于险种以及保险价格的谈判与约定，承包商与分包商之间，上述各主体在保险事故发生后的协商解决等。

一、保险合同

保险合同是投保人与保险人约定保险权利义务关系的协议。投保人是指与保险人订立保险合同，并按照合同约定负有支付保险费义务的人。保险人是指与投保人订立保险合同，并按照合同约定承担赔偿或者给付保险金责任的保险公司。根据保险合同，保险人有义务对投保人在保险范围内提出的索赔给予及时的理赔。保险人的义务主要体现在以下三点：第一，赔偿和给付保险金的义务；第二，及时签单的义务；第三，保密义务。

保险人赔偿或给付需满足以下条件：必须是保险标的受到损失；财产损失或人身灾害必须是保险合同中规定的危险引起的；财产保险损失的赔偿不能超过保险金额；财产损失应当发生在合同约定的地点或范围；人身保险中保险金的给付以保险金额为准；保险人须承担投保人或被保险人为减少保险标的的损失而付出的施救费用、诉讼费用和理赔费用。

二、合同要求与法律要求

（一）合同要求

面对建筑企业可能发生的各种建筑风险，以及可以购买的保险种类的多样性，承包商如何确定所需保险的种类在现实中是个较为复杂的问题，这使承包商经常不知所措。根据国际工程惯例与实践，以及国际通用的建筑工程合同文本规定，雇主和承包商根据不同保险类别，各自承担相应的责任。根据美国 AIA 合同文本 Document A201（Appendix C）第 11 条规定，保护项目的财产险和保护雇主的责任险，可能由雇主，也可能由承包商承担。AIA 合同将保险分为三部分，即承包商责任保险、雇主责任保险和财产保险。与 FIDIC“新红皮书”相比，AIA 合同中雇主明显地要承担更多的办理保险、支付保费方面的义务。AIA 合同规定，雇主应按照合同总价以及由他人提供材料或安装设备的费用投保并持有财产保险，该保险中包括了雇主以及承包商、分包商的权益，并规定雇主如果不准备按照合同条款购买财产保险，应在开工前通知承包商，这样承包商可以自己投保，以保护承包商、分包商的利益。承包商将以工程变更令的形式向雇主收取该保险费用。相对而言，承包商责任保险的种类较少，其主要承担人身伤亡方面的保险。

根据FIDIC“新红皮书”第18.2款规定，保险方应为工程、永久设备、材料以及承包商的文件投保，该保险的最低限额应不少于全部复原成本，包括补偿拆除和移走废弃物以及专业服务费和利润。当获得保险公司提供的相关保险后，保险公司有义务办理保险证书，证明相关范围已经获得保险公司的保险。雇主或承包商也可据此在保险事故发生后，向保险公司索赔。

（二）法律要求

相关责任人购买相应的保险是法律要求的，因此，无论建筑施工合同条款是否约定，承包商必须考虑相关保险事宜。例如，劳工赔偿保险、机动车保险、失业险、社会保障保险等是依照法律必须进行办理的。当然，关于失业险和社会保障险是否应计入商业保险范畴尚存在一定争论。法律要求，承包商应对其不作为或委托他人的行为后果造成的损失承担责任。此外，主包商对分包商负有责任，也就是说，主包商对分包商给雇主造成的损失负有连带责任。因此，无论法律规定是否涉及了一些保险种类，承包商作为工程的承建组织方必须考虑和提供不同种类的保险，以应对其自身行为引起的或者分包商行为引起的风险。至于保险费用的支付分担问题，也是承包商在投标时应予充分考虑，并与雇主协商的。

三、施工期项目的财产保险

1. 承包商风险保险的一切险

建筑工程一切险是承保各类民用、工业和公用事业建筑工程项目，包括道路、水坝、桥梁、港埠等，在建造过程中因自然灾害或意外事故而引起的一切损失。建筑工程一切险的被保险人包括雇主、总承包商、分包商、雇主聘用的监理工程师、与工程有密切关系的单位或个人，如贷款银行或投资人等。建筑工程一切险适用于所有建筑工程，尤其是住宅、商业用房、医院、学校、剧院、工业厂房、电站、公路、铁路、飞机场、桥梁、船闸、大坝、隧道、排灌工程、水渠及港埠等。建筑工程一切险保险金额需要根据费率来计算。

2. 承包商风险保险的指定险

主要指对特定或特殊事项的保险。

（1）扩展保险条款。这个险种覆盖所有由于暴风、冰雹、爆炸、暴乱、民众骚乱、航空器、车辆和烟雾引起的财产直接损失。

（2）故意破坏和损坏他人财产。

（3）水责任保险。水责任保险针对意外排放、泄露水或蒸汽的外溢等造成的财产损失提供保险赔偿。这个险种也包括有缺欠的管道、屋顶和水箱，但是不包括喷水器泄漏、洪水等引起的财产损失。

（4）喷水器泄漏险。这个险种对由于泄漏、凝固和喷水器安装问题造成项目的直接损失进行保护。

3. 地震保险

4. 蒸汽锅炉和机械设备保险

承包商或雇主对建设中的锅炉的试验、调试或被用来为墙面和地面工程加热时购买该类保险。与其他财产保险不同，这类保险包含了责任保险，即对承包商使用锅炉过程中造成的伤害和损害提供保障。

5. 安装流动保险

这个险种是针对指定险和一切险设立，例如项目机械（加热和空调系统）从离开运输的船体到安装在项目上并调试完毕的过程中可购买这类保险。这类保险在被保险人的保险利益终止，或项目被雇主接受时终止。

四、承包商自有物的财产保险

1. 承包商自有建筑物的财产保险

该险种对承包商办公室、工棚、仓库和其中的个人财产进行保险。

2. 承包商设备保险

该险种通常也称为流动保险，是指对承包商的建筑设备，无论其位于什么位置，进行投保。

3. 货运卡车货物保险

该险种涵盖了承包商自己货车承运的从供应商到仓库或施工现场材料和物资的指定风险的损失。

4. 运输流动险

该险种涵盖了属于承包商的财产或者其他人的财产在被公共运输公司运输时产生的风险。

5. 忠诚保险

该险种为承包商提供了由于其雇员不忠诚给承包商带来损失的补偿。

6. 犯罪保险

该险种用于保护承包商金钱、办公设备、证券和其他贵重物品因盗窃、抢劫、毁灭、消失等产生的损失。它也承担贵重物品因保险箱失窃或伪造产生的损失。

7. 贵重文件损毁保险

该险种用于保护承包商贵重文件的遗失、损毁，这些文件包括账簿、记录、地图、图纸、契约、抵押文件、合同和其他文件。但是此险种不保护误放、不能解释的消失、磨损、变质、虫蛀或战争引起的文件损失。

五、责任保险

1. 承包商的公共责任和财产损毁保险

该险种用于保护承包商伤害他人的行为所造成的损失或后果，即承包商对第三人的责任或者损毁他人财产的责任。

2. 承包商保护公共财产损失责任险

该险种适用于法律规定的由于分包商行为或不作为而造成的损失的保护。

3. 完工后的责任险

该险种用于已经完成或已经移交给雇主的工程项目，因承包商原因带来的项目瑕疵而可能给承包商带来损失的保险。雇主一般要求承包商承担此类责任，因为，一般保险通常只在完工前对承包商的工程建设设定保障，而很少在完工后对工程瑕疵提供保险。

4. 合同责任保险

该险种用于合同一方当事人，根据合同条款向另一方当事人承诺法律责任，未履行该责任而进行的保险。一般保险条款并不覆盖这种情形。

5. 职业责任保险

该险种用于承包商对雇主承诺的设计或其他职业服务责任的保险。

6. 雇员赔偿保险

该险种用于承包商对因公死亡或伤害的雇员进行赔偿的保险。

7. 雇主责任保险

该险种通常包含在雇员赔偿责任保险里，为承包商的雇员在雇用期间伤害和死亡提供了更广泛的保障，但是这个保险的承保范围和条件雇员赔偿法律的相关规定并不相同。

8. 雇主保护责任险

该保险险种保护雇主免受主承包商和分包商行为引起的损失。

六、代位求偿权（Subrogation）

保险的代位求偿权（Right of Subrogation）是指保险人享有的代位行使被保险人向造成保险标的损害的第三人进行索赔求偿的权利。严格来说，保险代位求偿权包括两个方面，权利代位和物上代位。保险代位求偿权仅指其中的权利代位。建设工程合同体系十分复杂，建设工程合同的保险事项主要是雇主与主承包商之间就建设工程施工项目所涵盖的财产责任等与保险公司进行的约定。在财产责任险方面，保险公司一般会要求投保人或被保险人用保险公司的赔偿换取其对加害第三人的代求偿权。这种权利要求，在一般的民事保险中并没有给当事人带来不便或利益损失，但是在建设工程保险中，由于分包商与主承包商对雇主的连带责任关系，使得由于分包商原因给雇主带来的损失由主承包商来承担。因此，如果主承包商或雇主的损失是由分包商造成的，在保险公司向雇主或主承包商赔偿之后，通过代位权向分包商追偿，最终分割的还是主承包商的利益，因此，在建设工程项目责任险方面，要求保险公司放弃部分责任险的代位求偿权是责任保险的难点和争议点。

第二节　保　险　信　函

Letter 1

To engineer, seeking approval to the names of insurers for employer's liability

Dear

In accordance with the provisions of clause ____, and to avoid any delays to the progress of the Works, we request the employer's approval by [*insert date*] to the following insurers to provide clause ____ insurance: [*insert name and address of the insurers*].

[*Or*:]

In accordance with the provisions of Schedule ____, paragraph ____, and to avoid any delays to the progress of the Works, we request the employer's approval by [*insert date*] to the following insurers to provide paragraph ____ insurance: [*insert name and address of the insurers*].

[*Or*:]

In accordance with the provisions of clause ____ and Schedule ____, paragraph ____, and to avoid any delays to the progress of the Works, we request the employer's approval by [*insert date*] to the following insurers to provide clause ____ and paragraph ____ insurance respectively: [*insert name and address of the insurers*].

Yours faithfully

信函 1

致工程师，要求批准雇主责任险承保方

敬启者：

根据合同第____条的规定，为了避免工程进度延误，我方要求雇主在［插入日期］前批准以下承保方提供合同第____条规定的保险：［插入承保方名称和地址］

［或者：］

根据____段，附表____的规定，为避免工程进度延误，我方要求雇主在［插入日期］前批准以下承保方提供____段中规定的保险：［插入承保方名称和地址］。

［或者：］

根据第____条和____段，附表____的规定，为避免工程进展延误，我方要求雇主在［插入日期］前批准承保方分别提供第____条和____段规定的保险：［插入承保方的名称和地址］。

敬上

Letter 2

To engineer, regarding contractor's insurance (a)

Special delivery

Dear

Thank you for your letter of the [*insert date*]. In response to your requirement, we have pleasure in enclosing documentary evidence in accordance with clause ____ of the contract for inspection by the employer. Please confirm that it is to the employer's satisfaction.

[*If appropriate, add either*:]

In accordance with Schedule ____, paragraph ____, we hereby submit for deposit with the employer the joint names policy referred to in paragraph ____.

[*Or*:]

In accordance with Schedule ____, paragraph ____, we enclose documentary evidence that we maintain appropriate All Risks cover independently of our obligations under this contract. Please note, for insertion in the appendix, that the annual renewal date is [*insert date*].

Yours faithfully

信函 2

致工程师，关于承包商保险（a）

快递

敬启者：

感谢贵方［插入日期］来函。按照贵方要求，我方很荣幸根据合同____项条款要求附上书面证明，供雇主审查。雇主对此满意，请予确认。

［如果合适，可补充：］

根据____段，附表____要求，我方特此向雇主提交____段落中涉及的联名保险抵押金。

［或：］

根据第____段，附表____的要求，我方附上书面证明，我方就本合同项下的我方责任单独上全险。请注意每年续保日期为［插入日期］，插入附录。

敬上

Letter 3

To engineer, regarding contractor's insurance (b)
Special delivery

Dear

Thank you for your letter of the [*insert date*]. In response to your requirement, we have pleasure in enclosing documentary evidence that the insurances required under clause ____ have been taken out and are in force at all material times.

[*If appropriate, add either*:]
We also enclose documentary evidence that the insurance required under clause ____ of the contract has been taken out and is in force at all material times.

[*Or*:]
We enclose documentary evidence that we maintain appropriate All Risks cover independently of our obligations under this contract.

[*Then add*:]
Please confirm that these insurances are all to the employer's satisfaction.

Yours faithfully

信函 3

致工程师，关于承包商保险（b）
快递

敬启者：

感谢贵方［插入日期］来函。按照贵方要求，我方很荣幸附上书面证明，我方已购买合同第____条所要求的保险，并在此期间有效。

［如果合适，可补充：］
我方同样附上书面证明，我方已购买合同第____条所要求的保险，并在此期间有效。

［或者：］
我方附上书面证明，我方就本合同项下的我方责任单独上全险。

［然后补充：］
请确认这些保险满足雇主要求。

敬上

Letter 4

To employer, regarding contractor's insurance (c)

Special delivery

Dear

Thank you for your letter of the [*insert date*]. In response to your request, we have pleasure in enclosing copies of the following insurance policies in accordance with clause ____ of the conditions of contract.

[*List policies together with numbers and dates*]

Yours faithfully

信函 4

致雇主，关于承包商保险（c）

快递

敬启者：

感谢贵方［插入日期］来函。按照贵方要求，我方很高兴根据合同条款第____条的要求附上以下保险单副本。

［列出保单编号与日期］

敬上

Letter 5

To employer, within 21 days of acceptance of tender or renewal of insurance

Special delivery

Dear

In accordance with clause ____ of the conditions of contract, we have pleasure in enclosing a certificate from our insurer/broker [*delete as appropriate*] attesting that the appropriate insurance policies have been effected.

Yours faithfully

信函 5

致雇主，在接受投标或保险续保 21 天内

快递

敬启者：

依据合同条款____项规定，我方很高兴附上我方承保方/经纪人［酌情删除］出具的保险凭证，证明相关保单已经生效。

敬上

Letter 6

To engineer, after approval of insurers for employer's liability

Special delivery

Dear

We enclose insurance policy number [*insert number*] and premium receipt dated [*insert date*] in respect of insurance taken out with approved insurers, [*insert name*], for deposit with the employer in accordance with clause ____ in joint names.

Yours faithfully

信函 6

致工程师，承保方批准雇主责任险后

快递

敬启者：

随函附上在认可的承保方［插入名称］处所投保险的保险单［插入编号］和［插入日期］保费收据，要求雇主依据条款____规定联名支付保证金。

敬上

Letter 7

To engineer, regarding professional indemnity insurance

Dear

Further to your letter dated [*insert date*] requesting documentary evidence that the insurance required under clause ____ is being maintained, we have pleasure in enclosing the following documents [*list*].

[*If any original documents have been included, send the letter by special delivery and add*:]

Please note that we have included the originals of documents [*identify*] and we should be grateful if you would return them, using special delivery, by [*insert date*].

Yours faithfully

信函 7

致工程师，关于职业责任保险

敬启者：

[插入日期] 贵方来函要求提供书面证据，证明合同条款____所要求的保险一直有效，很高兴附上下列文件 [如下]。

[如含任何原件，烦以快递发函，请补充：]

望贵方注意，我方已提交原件 [确认]，若贵方能在 [插入日期] 前以快递返还文件，将不胜感激。

敬上

Letter 8

To engineer, if professional indemnity insurance is no longer available at commercially reasonable rates

Dear

In attempting to renew our professional indemnity insurance, we have to inform you in accordance with clause ____ that the insurance referred to in clause ____ is no longer available at commercially reasonable rates.

We should be glad to discuss the matter with you if you will let us know a suitable time and date, bearing in mind that the insurance renewal date is [*insert date*].

Yours faithfully

信函 8

致工程师，如果不能再以合理的商业费率购买职业责任保险

敬启者：

我方试图续保职业责任保险，依据条款____，我方必须通知贵方我方无法再以合理的商业费率购买条款____中所涉及的保险。

如果贵方告知我方合适的时间与日期，我方愿与贵方讨论此事。此外，请注意续保日期为［插入日期］。

敬上

Letter 9

To engineer, if Joint Fire Code remedial measures are a variation

Dear

We hereby give notice as required by clause ____ that the Remedial Measures as required by the insurers under the provisions of clause ____ require a variation.

Please issue such instructions as are necessary to enable compliance.

Please take this letter as a notice of delay under clause ____ involving a Relevant Event under clause ____ and an application for loss and/or expense under clause ____ involving a Relevant Matter under clause ____.

Yours faithfully

信函 9

致工程师，如果《联合消防规范》补救措施构成施工变更

敬启者：

根据条款____要求，我方特此通知贵方承保方根据条款____所要求的补救措施需要更改施工。

请贵方发布必要的指示以便其能符合规范。

根据条款____的要求，该条款包含条款____中规定的相关事项，我方发出延期通知。此外，根据条款____的要求，该条款包含条款____中规定的相关事项，我方申请损失和/或费用赔偿。

敬上

Letter 10

To engineer, if Joint Fire Code remedial measures are a variation, but require emergency action

Dear

We hereby give notice as required by clause ____ that we have been obliged to commence the carrying out of work constituting emergency compliance with Remedial Measures as required by the insurers under the provisions of clause ____. The steps we are taking to supply limited materials and execute limited work are as follows:

[*Insert details.*]

We confirm that the work carried out and the materials supplied rank as a variation. Please provide further instructions necessary.

Please take this letter as a notice of delay under clause ____ involving a Relevant Event under clause ____ and an application for loss and/or expense under clause ____ involving a Relevant Matter under clause ____.

Yours faithfully

信函 10

致工程师，如果《联合消防规范》补救措施构成变更，但需采取紧急措施

敬启者：

根据条款____的要求，我方特此通知贵方我方必须开始实施承保方依据合同条款____要求的符合补救措施的紧急施工工作。我方正在采取措施提供有限的材料执行有限的工作，具体如下：

[插入细节。]

我方确认所做的工作和提供的材料构成施工变更，请贵方提供进一步指示。

根据条款____的要求，该条款包含条款____中规定的相关事项，我方发出延期通知。此外，根据条款____的要求，该条款包含条款____中规定的相关事项，我方申请损失和/或费用赔偿。

敬上

Letter 11

To employer, regarding employer's insurance

Dear

We should be pleased to receive documentary evidence and receipts showing that the joint names insurance policy has been taken out and is being maintained in accordance with ____ [*delete as appropriate*] of the contract. The receipt should indicate that such insurance is currently effective and it should be in our hands by [*insert date*].

Yours faithfully

信函 11

致雇主，关于雇主险

敬启者：

我方希望收到书面证明与收据，以此证明贵方已经依据合同____［酌情删除］要求投保了联名险且该保险一直有效。该收据应体现该保险现在有效并请于［插入日期］前将其交给我方。

敬上

Letter 12

To employer, who fails to maintain insurance cover (a)

Dear

We refer to your telephone conversation with our M... [*insert name*] today and we confirm that you are unable to produce a receipt, as required by ____ [*delete as appropriate*], showing that insurance is currently effective.

In view of the importance of the insurance and without prejudice to your liabilities, we are arranging to exercise our rights under the above mentioned clause immediately. On production of a receipt for any premium paid, we will be entitled under ____ [*delete as appropriate*] to have its amount added to the contract sum.

[If a failure to insure under ____, add:]

We draw your attention to the fact that we intend to exercise our right of entry and inspection for the purposes of survey and inventory on [*insert date*] at approximately [*insert time*].

Yours faithfully

Copy: Engineer

信函 12

致雇主，没有续保（a）

敬启者：

我方［插入姓名］经理今日与贵方通话，确认贵方未能按照合同____［酌情删除］的要求出示收据以证明保险现在有效。

鉴于保险的重要性以及在不损害贵方责任的前提下，我方正在安排按照上述条款行使我方权利。关于任何已付保费收据，我方有权根据____条款的规定［酌情删除］将其金额追加至合同总金额中。

［如果未能依据____项投保，请补充：］

我方提请贵方注意我方打算于［插入日期］大约［插入时间］行使我方入场检查权对贵方进行调查。

敬上

抄送至：工程师

Letter 13

To employer，who fails to maintain insurance cover (b)

Dear

We refer to your telephone conversation with our M... [*insert name*] today and we confirm that you are unable to produce evidence，as required by clause ____，that the insurance referred to in clause ____ [*delete as appropriate*] has been taken out and is in force at all material times.

You are in breach of contract for which we intend to claim damages. We are advised that such damages include our costs in taking out appropriate insurance.

Yours faithfully

信函 13

致雇主，没有续保（b）

敬启者：

我方［插入姓名］经理今日与贵方通话，确认贵方未能按照合同____［酌情删除］的要求出示证据以证明贵方已经购买了合同条款____所涉及的保险且该保险出险时有效。

贵方构成违约，我方打算为此要求索赔，我方获悉该损失包括投保费用。

敬上

Letter 14

To engineer and employer, if any damage occurs due to an insured risk

Dear

In accordance with ____ [*delete as appropriate*] of the conditions of contract, we hereby give notice that loss/damage [*delete as appropriate*] has been caused to the work executed/site materials [*delete as appropriate*] by one of the risks covered by the joint names policy.

[*Describe the loss or damage, stating extent, nature and location.*]

Yours faithfully

信函 14

致工程师和雇主，如果保险事故造成损失

敬启者：

按照合同____［酌情删除］的要求，我方特此通知贵方工程/施工材料［酌情删除］出现了联名保单覆盖的保险事故，造成了损失/损害［酌情删除］。

［描述所受损失或损害，说明其程度，性质及出险位置。］

敬上

第五章 现场管理信函

第一节 现场管理概述

一、施工现场管理含义

国际工程现场管理是指承包商或项目部依照项目所在国政府部门的建筑施工管理规定、企业内部的施工现场管理规定等内容，对工程施工项目现场以内、工程主体以外所包含的内容进行的管理，主要包括现场平面管理、材料设备管理、环境管理、安全管理、人员管理等。

国际工程施工现场管理涉及多方面内容，为了保证现场施工工作的有效进行，施工现场不仅要做好平面布置、设备管理和材料管理，还要根据当地的实际情况进行严格的安全管理和环境管理。由于施工现场包含大量不同国家的人员，导致人员的管理更加复杂，所以国际工程施工现场的安全管理不仅包括对施工区域的安全管理，也包括对大面积工人生活区的安全管理。

由于工程项目建设地点位于国外，许多因素都会对工程施工现场管理造成影响，主要包括：1）法律因素，不同法系国家的法律制度差异较大，对法律的理解也不尽相同；2）气候、经济因素；3）宗教信仰、行为习惯因素。在亚洲和非洲地区的施工现场，员工的生活习惯和宗教信仰差距较大，特别是人员组成复杂，语言不通成了影响施工现场管理的重要因素。

因此，在国际工程管理过程中，除非项目规模很小，否则承包商撰写大量信函与雇主、工程师和分包商进行沟通协同，并通过信函解决相关问题是不可避免的。例如关于工程总进度计划、现场占用、图纸、现场会议、工程师指示以及缺欠材料等事宜都需要以信函形式进行沟通。

二、FIDIC“新红皮书”关于雇主提供承包商现场占有的规定

提供施工现场是雇主的义务。“进入现场”（access to site）具有多重含义，可以指进入现场区域的方式和可能性，也可指占有现场的能力。进入现场意味着允许承包商的人员、设备、材料、车辆、服务能够到达现场区域，履行施工合同项下的义务。为了有效进入现场，应允许使用各种运输方式，并且雇主必须准备妥当，使现场具备进入的条件。

FIDIC“新红皮书”第2.1款“现场进入权”规定：“雇主应在投标附录中规定的时间（或几个时间）内，给予承包商进入和占有现场所有部分的权力。承包商不能独享此项进入和占有权”。根据合同规定，如果要求雇主（向承包商）给予任何基础、结构、生产设备或进入手段的占有权，雇主应按规范规定的时间和方式给予。同时，无论施工合同是否以明示方式规定了雇主提供现场进入和占有权的条款，施工合同都存在一项默示条款，即雇主应在合理的时间内向承包商提供现场占有权，以便承包商履行合同的权利和义务，

按期完成工程项目。因此，在缺少明示规定时，施工合同中通常默示一项条款，即雇主应在合理的时间内向承包商移交现场，在大多数情况下，应向承包商提供充分的不受干扰的占有权，以使承包商能够履行合同项下的义务。

然而，如果雇主未能按照合同规定或双方商定的时间和方式提供现场进入权和占有权，将对项目产生重大影响。在 FIDIC 合同条件下，这将会导致一系列时间的改变，也会导致竣工日期的延误和承包商费用的增加，还会导致承包商不得不修改根据第 8.3 款“进度计划”的规定提交的进度计划和施工次序。同时，如果雇主未能按照合同规定或双方商定的时间和方式提供现场进入权和占有权将不可避免地产生两个后果，一是施工进度的延误，二是承包商会遭受损失和发生额外费用。对此，有些标准合同文件规定了承包商有权索赔工期和要求雇主给予费用补偿，但有些合同没有明示的规定。在缺少明示条款规定的情况下，施工合同默示承包商有权提出工期延长索赔和要求雇主给予费用补偿。例如，按照 FIDIC“新红皮书”第 2.1 款的明示规定，如果雇主未能提供现场进入权和占有权，则承包商有权提出工期延长索赔和费用索赔。

第二节 现场管理信函

Letter 1

To employer, if possession not given on the due or the deferred date

Special delivery

Dear

[*If deferment of possession clause does not apply*:]

Possession of the site should have been given to us on [insert date] in accordance with clause ____ of the conditions of contract. Possession was not given to us on the due date.

[*If deferment of possession clause does apply, but deferred date not met*:]

Under clause ____ you deferred the giving of possession until [insert date]. You did not give us possession on the deferred date.

[*Then either*:]

Moreover, the engineer has informed us today by telephone that the date for possession remains uncertain.

[*Or*:]

The engineer has informed us today by telephone that you will be unable to give possession until [*insert date*].

[*continued*]

信函 1

致雇主，如果没在到期日或递延日交付施工现场

快递

敬启者：

[如果延期交付条款不适用：]

根据合同条款第____条，贵方应在[插入日期]向我方交付施工现场。贵方在到期日没有交付。

[如果延期交付条款适用，但延期日期不符：]

根据合同条款第____条，贵方推迟交付日期到[插入日期]。贵方在延期日期没有交付。

[然后或者：]

此外，工程师今天电话通知我方，交付日期仍然不确定。

[或者：]

工程师今天电话通知我方，贵方将无法在[插入日期]之前交付。

[待续]

Letter 1 continued

[*Then*:]

By copy of this letter to the engineer, we give notice of delay under clause ____ involving a Relevant Event under clause ____ and make application for loss and/or expense under clause ____ involving a Relevant Matter under clause ____ . However, such notice and application is without prejudice to any other rights and remedies including (but without limitation) the right to treat the failure to give possession as a serious breach of contract entitling us to damages and, if prolonged, to treat it as a repudiation of the contract on your part.

Naturally, we hope that your difficulties will be resolved and we should welcome a meeting with you to discuss the prospects of taking possession of the site. To that end we are generally available on [*insert list of dates*] and suggest that you telephone us to arrange a meeting.

Yours faithfully

Copy: Engineer

信函1 续:

[然后:]

将此函抄送给工程师，根据涉及相关事项的合同第____项的条款____内容，我方发出延期通知，并根据涉及相关事项的合同第____项的条款____内容，我方申请损失和/或费用赔偿。但是，此类通知和申请并不影响我方任何其他权利和补救措施，包括（但不限于）有权将未能让出现场视为严重违约行为，合同规定我方有权提出损害赔偿，如果拖延让出现场，我方有权将其视为贵方毁约。

当然，我方希望贵方的困难能够得到解决，我方欢迎与贵方进行会谈，讨论获得进场的可能性。为此，我方一般在[列出日期]有时间，并建议贵方通过电话与我方安排会议。

敬上

抄送至：工程师

Letter 2

To employer, if possession not given on the due date

Dear

Possession of the site should have been given to enable us to commence the works on [*insert date*] in accordance with clause _ of the conditions of contract. Possession was not given to us on the due date.

[*Then either*:]

The engineer has informed us today by telephone that the date when you will be able to give possession is still uncertain.

[*Or*:]

The engineer has informed us today by telephone that you will be unable to give possession until [*insert date*].

[*Then*:]

This is a serious breach of contract for which we will require appropriate compensation and we reserve all our rights and remedies in this matter. Without prejudice to the foregoing, we suggest that a meeting would be useful and look forward to hearing from you.

Yours faithfully

信函 2

致雇主，如果未按期交付施工场地

敬启者：

依据合同条款第____项，为使我方能于［插入日期］顺利开工，项目现场应于［插入日期］交至我方，但贵方并未如期交付。

［或者:］

工程师今日致电我方称贵方交付施工现场日期仍未确定。

［或者:］

工程师今日致电我方，贵方于［插入日期］才能够交付施工现场。

［然后:］

此行为已严重违反合同，我方将为此要求相应补偿，同时，我方对此保留所有权利和补救措施。在不影响上述情况下，我方认为有必要进行开会讨论，期待尽快收到贵方答复。

敬上

Letter 3
To engineer, if date for possession advanced

Dear

Thank you for your letter of the [*insert date*] advising us that the employer is prepared to allow us to take possession of the site on [*insert date*].

Naturally, we will endeavour to take advantage of the opportunity to make an earlier start than we anticipated and we will inform you when we intend to take possession. There are many things to take into consideration and some reorganising of our labour resources would be involved, all of which would result in additional costs, which we would expect the employer to reimburse. Please let us know, by return, if the employer will reimburse such costs so that we can make our decision accordingly. If we take possession of the site on the date you suggest, such action will not affect our obligation to complete the works on the date for completion stated in the Contract Particulars.

Yours faithfully

信函 3
致工程师，如果交付施工现场日期提前

敬启者：

感谢贵方［插入日期］来函，告知雇主将于［插入日期］允许我方接管施工现场。

我方将努力借此机会提前开工，我方会通知贵方何时接管现场。为了提前进场施工，我方要考虑许多因素，需要重新组织劳动力资源，所有这些工作都会产生额外费用，希望雇主能给予补偿。如果雇主同意补偿这些费用，请速告知我方，以便我方可以作出相应决定。若我方按照贵方建议时间进场，此行为不会影响我方按照承包商合同规定的竣工日期完成工程的义务。

敬上

Letter 4

To engineer，regarding items in minutes of site meeting

Dear

We have examined the minutes of the meeting held on the [*insert date*]，which we received today. We have the following comments to make：

[*List comments*.]

Please arrange to have these comments published at the next meeting and inserted in the appropriate place in the minutes.

Yours faithfully

Copies：All present at meeting and those included in the original circulation

信函 4

致工程师，关于现场会议纪要的记录事项

敬启者：

我方今日收到并查阅了［插入日期］的会议纪要。我方意见如下：

［列出具体意见。］

请安排在下次会议上公布这些意见，并添加在会议纪要中的适当位置。

敬上

抄送：所有与会人员及原来的相关人员

Letter 5

To engineer, requesting information

Dear

We should be pleased to receive the following information which it is necessary for us to receive by the dates stated in order to enable us to carry out and complete the Works in accordance with the conditions of contract.

[*List the information required and the date by which each item of information must be received. Wherever possible, allow the engineer at least* 14 *days to prepare the information.*]

Yours faithfully

信函 5

致工程师，要求提供信息

敬启者：

为了能够按照合同规定实施并完成工程，如能在规定日期前收到以下关键信息，我方将十分欣喜。

[列出所需信息，以及提供每项信息的截止日期，如果条件允许，应给工程师留出至少两周时间来准备以上信息。]

敬上

Letter 6

To engineer, if insufficient information on setting out drawings

Dear

We are preparing to commence work on site on [*insert date*]. Our first task will be to set out the Works. An examination of the drawings you have supplied to us reveals that there is insufficient information for us to set out the Works accurately. We enclose a copy of your drawing number [*insert number*] on which we have indicated in red the positions where we need dimensions/levels [*delete as appropriate*].

We need this information by [*insert date*] in order to avoid delay and disruption to the Works.

Yours faithfully

信函 6

致工程师，如果放样图的信息不充分

敬启者：

我方准备于［插入日期］开工。我方的首要任务是做好施工放线。我方检查了贵方所提供的图纸，图纸信息不充分，我方无法准确地进行施工放线。我方随函附上贵方图纸一份，编号为［插入号码］。需要注明尺寸/高度的部分［酌情删除］，已用红色标注。

为避免工程延误或中断，我方需于［插入日期］前收到以上信息。

敬上

Letter 7

To engineer, if information received late

Dear

On [*insert date*] we requested the following drawings/details/instructions [*delete as appropriate*] by [*insert date*]. They have not been received.

Clause ____ requires you to provide such information 'as and when from time to time may be necessary'. You are clearly in breach of your obligations. We have complied with our duty in the last part of the clause, to advise you sufficiently in advance of the time when it is necessary for us to receive the information. The absence of the information is causing us delay and disruption. Please inform us, by return, when we can expect it to arrive.

Take this as notice of delay under clause ____ and an application for loss and/or expense under clause ____. We expect to be able to provide further details in due course.

Yours faithfully

信函 7

致工程师，如果延迟收到信息

敬启者：

我方已于［插入日期］要求贵方在［插入日期］前提供下列图纸/详细信息/指示［酌情删除］。但我方仍未收到。

合同条款____要求贵方在“必要时随时”提供以上信息。显然贵方并没有履行此义务。我方已遵守该条款最后部分的我方职责，提前给贵方充分时间提供该信息。此类信息的缺失会导致工程的延迟与中断。请贵方立即回复我方何时能够收到此信息。

依照合同条款第____项，贵方可视此为我方发出的延期通知，并且根据条款第____项，我方申请贵方赔付我方相应的损失/费用。我方期望能够在适当时候提供进一步详情。

敬上

Letter 8

To engineer, if information not received in accordance with the information release schedule

Dear

We draw your attention to the information release schedule for this project which indicates that [*describe the information and the date on which it should have been provided*]. It has not been received.

You are clearly in breach of your obligations as set out in clause ____. The absence of the information is causing us delay and disruption. Please inform us, by return, when we can expect it to arrive.

Take this as notice of delay under clause ____ and an application for loss and/or expense under clause ____. We expect to be able to provide further details in due course.

Yours faithfully

信函 8

致工程师，如未按照信息发布时间表收到信息

敬启者：

我方希望贵方注意该项目的信息发布时间表，表中包含［描述信息和需要提供信息的时间］。但目前我方还未收到相关信息。

显然，贵方没能履行合同条款第____项所规定的贵方义务。缺失此类信息正导致我方工程延迟与中断。请贵方尽快回复我方何时能够收到此信息。

依照合同条款第____项，贵方可视此为我方发出的延期通知，并且根据条款第____项，我方申请贵方赔付我方相应的损失/费用。我方期望能够在适当时候提供进一步详情。

敬上

Letter 9

To engineer, if contractor providing contractor's design documents

Special delivery

Dear

In accordance with the submission procedure in Schedule ____, we enclose two copies of the following contractor's design documents:

[*List drawings, specifications, details and, if requested, calculations.*]

Please return the documents within 14 days of receipt, marked 'A', 'B' or 'C'. Please note that where you return a document marked 'B' or 'C', paragraph ____ requires you to identify in writing why you consider it is not in accordance with the contract.

Any documents not returned within 14 days will be regarded as marked 'A' in accordance with paragraph ____.

Yours faithfully

信函 9

致工程师，如果承包商提供承包商设计文件

快递

敬启者：

按照附表____的提交程序要求，我方附上如下两份承包商设计文件。

[列出图纸，规范，细节，如有要求，列出具体计算信息。]

请在收到后 14 天内返还文件，标明“A”，“B”或“C”。请注意，在贵方返还标有“B”或“C”的文件时，第____段要求贵方以书面形式说明贵方认为不符合合同要求的原因。

按照合同第____段规定，任何未在 14 天内退回的文件将被视为标记为“A”。

敬上

Letter 10

To engineer, who fails to return the contractor's drawings in due time

Dear

We refer to our letter of the [*insert date*] enclosing two copies of drawings numbered [*insert numbers*], being the contractor's design documents that we submitted under paragraph ____ of the Schedule 1 design submission procedure. Under paragraph ____, you should have responded within 14 days of receipt of our submission. At the date of this letter, we have not received such response.

Paragraph ____ provides that, in these circumstances, the design documents will be regarded as marked 'A' and we are proceeding accordingly. Please take note that if you subsequently return the documents with comments attached, we will quite properly consider them to be instructions requiring variations to the Works and, as well as the cost of such variations, we shall be entitled to extension of time and loss and/or expense.

Yours faithfully

信函 10

致工程师，未按期归还承包商图纸

敬启者：

我方于［插入日期］发函贵方，并且依照设计提交流程附表 1，第____段内容要求，随函附上编号为［插入编号］的图纸，即承包商的设计文件。依照第____段内容要求，贵方本应在收到我方信函后 14 日内作出答复，但我方目前仍未收到此回复。

第____段内容规定，在此情况下，设计文件将被视为“A”类文件，我方正按此施工。请注意，如果贵方在此之后回函，并附上相关意见，我方将视其为要求变更工程的指示，以及此工程变动所产生的费用。我方将有权延长工期，并向贵方追索损失/费用赔偿。

敬上

Letter 11

To engineer, if contractor providing contractor's design documents

Actual/special delivery

Dear

In accordance with clause ____ of the conditions of contract, we enclose two copies of the following contractor's design documents:

[*List drawings, specifications, details and, if requested, calculations.*]

[*If there is a submission procedure stated in the contract, follow it, if there is no procedure, add:*]

Please confirm within 7 days of receipt that we may proceed with construction using these documents. Bear in mind that if you instruct us to amend any of the documents, such amendment will be a variation to be valued under section ____ of the contract unless such documents can be demonstrated to be not in accordance with the contract. If you do not respond within 7 days, we shall proceed on the basis that you have approved the documents.

Yours faithfully

信函 11

致工程师，如果承包商提供其设计文件

亲自交付/快递

敬启者：

根据合同条款第____项，我方附上以下两份承包商设计文件的副本。具体信息如下：

[列出图纸，规范，细节，如有要求，列出具体计算信息。]

[如果合同中有提交程序，请遵循，如果没有此项程序，请补充：]

请贵方在收到文件 7 日内确认，我方可以使用这些文件继续施工。请记住，如果贵方指示我方修改任何文件，根据合同第____部分内容，该修改将构成估价变化，除非贵方可以证明该文件与合同不符。如果贵方没有在 7 日内作出答复，我方将认为贵方已经批准该文件并继续施工。

敬上

Letter 12

To engineer, if contractor providing levels and setting out information

Actual/special delivery

Dear

In accordance with clause ____ of the conditions of contract, we enclose two copies of the levels and setting out dimensions we propose to use for the Works.

In order to avoid any delays, please let us know by close of business on [*insert the latest date you need to know, allowing for any changes that may need to be made after notification*] if you have any adverse comments. If we do not receive any comments, we shall assume that you approve the information and proceed to use it in constructing the Works.

Yours faithfully

信函 12

致工程师，如果承包商提供标高和放线信息

亲自交付/快递

敬启者：

根据合同条款第____项，我方附上两份文件，建议该工程使用的标高和放线范围。

为了避免工期延误，如果贵方对此有相反意见，请在下班前告知我方［插入你方可接受的最晚日期，以便在接到通知后进行调整］。如我方未收到任何意见，将假定贵方已批准该数据，并继续使用该数据施工。

敬上

Letter 13

To engineer, who returns contractor's drawings with comments

Dear

Thank you for your letter of the [*insert date*] with which you returned our drawings numbers [*insert numbers*] with comments.

Although we are always grateful for any comments you feel able to make, we should point out that the purpose of clause ___ of the conditions of contract is to provide you with copies of the drawings and other information we intend to use to carry out the work. The drawings have been prepared strictly in accordance with the Employer's Requirements.

If, therefore, you intend us to act upon your comments, please issue them as instructions requiring a variation under the provisions of clause ____ . If we have no reasonable objection, such an instruction would be subject to valuation under clause ____, extension of time under clause ____ and loss and/or expense under clause ____ .

Yours faithfully

信函 13

致工程师，返还承包商图纸并给出意见

敬启者：

感谢贵方［插入日期］来函，并已收到随函附上的带有贵方修改意见的图纸［插入图纸编号］。

对于贵方所提的任何意见，我方一直很感激，但我方应该指出，合同条款第____条的目的是规定我方向贵方提供我方打算用来开展施工的图纸和其他信息的副本。图纸是严格按照雇主要求准备的。

因此，如果贵方打算要求我方按照贵方修改意见施工，根据合同条款____项规定，请将其作为要求变更施工的指示正式发出。若我方无合理反对意见，根据合同条款____，该指示会导致计价；根据条款____，会导致工期延长；根据条款____，会导致损失以及费用赔偿问题。

敬上

Letter 14

To engineer, if discrepancy found between documents

Dear

In accordance with clause ____ we bring to your attention the following discrepancies which we have discovered:

[*List, giving precise details of bills of quantities, specification or work schedule references, drawing numbers or dates and numbers of engineer's instructions.*]

In order to avoid delay or disruption to our progress, we require your instructions by [*insert date*].

Yours faithfully

信函 14

致工程师，如果两个文件之间出现不符的情况

敬启者：

按照合同条款____，我方提请贵方注意我方发现如下不符之处：

[列出，提供以下事项的详细信息：工程量清单，规范或工程进度表参考，图纸编号或工程师指示的编号与日期。]

为避免工程延误或中断，我方要求贵方于［插入日期］前发出指示。

敬上

Letter 15

To engineer, if discrepancy within the Employer's Requirements

Dear

We have found a discrepancy in the Employer's Requirements [*give details*].
Our Contractor's Proposals do not deal with the matter and, therefore, we propose the following amendment:

[*Describe in detail how to deal with the discrepancy including the provision of any drawings.*]

Please either let us have your written agreement to our proposal or details of your alternative decision, either of which will rank as a change to the Employer's Requirements. We need your agreement or decision by [*insert date*] in order to avoid delay to the Works.

Yours faithfully

信函 15

致工程师，如果“雇主要求”存在矛盾

敬启者：

我方在“雇主要求”中发现了相互矛盾之处［列举出细节］。我方的“承包商建议书”不处理此问题。因此，我方建议作出以下修改：

［请详细描述如何处理这种矛盾，包括提供图纸。］

请贵方为我方要么提供书面同意书，同意我方建议，要么提供贵方决策的详情。其中任何一种情况均将变更“雇主要求”。为避免工程延误，我方需要于［插入日期］前获得贵方同意书或决策意见。

敬上

Letter 16

To engineer, if discrepancy within the Contractor's Proposals

Dear

We have found a discrepancy in our Contractor's Proposals [*give details*]. We propose the following amendment to remove the discrepancy:

[*Describe in detail how to deal with the discrepancy including the provision of any drawings.*]

In accordance with clause ____ of the conditions of contract, please either decide which of the discrepant items you prefer or accept our proposed amendment. We require your decision or acceptance in writing by [*insert date*] in order to avoid delay to the Works.

Yours faithfully

信函 16

致工程师，如果“承包商建议书”中存在矛盾

敬启者：

我方在“承包商建议书中”发现自相矛盾之处［列举出细节］。为消除此矛盾之处，我方建议作出以下修订：

［请详细描述如何处理这种矛盾，包括提供图纸。］

按照合同条款____，请贵方要么选择贵方偏好的事项，要么接受我方提出的修改意见。为避免工程延误，我方要求贵方于［插入日期］前作出书面决定或书面接受我方的修改意见。

敬上

Letter 17

To engineer, if discrepancy found between Employer's Requirements and Contractor's Proposals

Dear

We have found a discrepancy between the Employer's Requirements and our Contractor's Proposals [*give details*]. The contract does not expressly deal with this situation. If you are content to accept the way we have dealt with the matter in our Contractor's Proposals, there is no problem, otherwise we suggest that a meeting on site is required. Please let us have your response as soon as possible/as a matter of urgency [*delete as appropriate*].

Yours faithfully

信函 17

致工程师，如果“雇主要求”和“承包商建议书”之间存在矛盾

敬启者：

我方发现“雇主要求”和“承包商建议书”之间存在矛盾［详细说明］。合同条款没有明确处理这种情况。如果贵方接受我方按“承包商建议书”要求处理该事，没有问题。否则，我方建议有必要召开现场会议。请贵方尽快回复/将其视为紧急事项处理［酌情删除］。

敬上

Letter 18

To engineer, if alleging that contractor should have checked the design

Dear

We are in receipt of your letter dated [*insert date*] in which you assert that we are responsible for the design of [*insert the element concerned*] or at the very least for checking it. This element is a design which you prepared and forms part of the Employer's Requirements.

Our obligation under clause ____ is to complete the design of the Contractor's designed portion. This duty is clarified by clause ____ which states that we are not responsible for the contents of the Employer's Requirements nor for verifying the accuracy of any design contained within them.

Therefore, we are neither responsible for the design to which you refer nor for checking that it works.

Yours faithfully

信函 18

致工程师，如果宣称承包商应该检查设计

敬启者：

我方已收到贵方［插入日期］来函，贵方在函中声称我方应负责设计［插入有关部分］或至少对其进行检查。该部分的设计由贵方负责，这是“雇主要求”中的一项内容。

根据条款第____项规定，我方的义务是完成承包商设计部分的设计工作。该义务在条款____中有明确规定，该条款明确说明我方不对“雇主要求”中的内容负责，也不负责核实其中包含的任何设计的准确性。

因此，我方既不负责贵方提及的设计也不负责对其进行检查。

敬上

Letter 19

To employer，pointing out design error in Employer's Requirements

Dear

In carrying out the detail design for which we are responsible，we have discovered that part of the design which you prepared and which forms part of the Employer's Requirements appears to be defective in that：

［*Briefly indicate the nature of the defective design.*］

Our obligation under clause ____ is merely to complete the design of the contractor's designed portion. This duty is clarified by clause ____ which expressly states that we are not responsible for the contents of the Employer's Requirements nor for verifying the accuracy of any design contained within them.

We believe that it is now a matter for you to issue a change instruction to vary the Employer's Requirements to enable us to proceed with our part of the design. Please be aware that this is now causing us a delay for which we shall expect an appropriate extension of time in due course. It may also give rise to direct loss and/or expense for which we will seek reimbursement as soon as the position becomes clear.

Yours faithfully

信函 19

致雇主，指出“雇主要求”中的设计错误

敬启者：

在执行我方负责的细节设计过程中，我方发现由贵方负责设计的部分存在缺陷，该部分设计是“雇主要求”中的事项。原因是：

［简要说明设计缺陷的性质。］

根据条款第____项要求，我方的义务是完成承包商设计部分的设计工作。该义务在条款____中有明确规定，该条款明确说明我方不对“雇主要求”中的内容负责，也不负责核实其中包含的任何设计的准确性。

为使我方能够继续我方的设计工作，我方认为现在应由贵方发出变更指示以更改“雇主要求”。请贵方注意，此事正造成延误，我方预计工期会适当延长。此事也会导致直接损失和/或费用，一旦情况明确，我方将索求赔偿。

敬上

Letter 20

To engineer, requesting directions to integrate the design

Dear

We refer to [*insert description of the element concerned*], which is part of the contractor's designed portion of the Works.

It is unclear how this element is to be integrated into the design of the Works as a whole. [*Add either*:] This was not clear in the invitation to tender documents. [*Or*:] This is due to the issue of your instruction number [*insert number*] dated [*insert date*].

Therefore, we request your instructions for integration under clause ____ by [*insert date*] so as to avoid delay and extra costs to the project.

Yours faithfully

信函 20

致工程师，要求提供指示，整合设计方案

敬启者：

今提及［插入有关部分的说明］，该部分为工程项目中承包商负责设计的部分。

我方还不清楚如何将该部分整合到工程整体设计中。［要么补充：］此情况在邀标文件中未做明确说明。［或：］该情况是由于贵方在［插入日期］的指示［插入编号］造成的。

因此，根据条款第____项，为避免造成项目延误或产生额外费用，我方要求贵方在［插入日期］前发布整合设计方案的指示。

敬上

Letter 21

To engineer, noting divergence between statutory requirements and other documents (a)

Dear

We have found what appears to be a divergence between statutory requirements and your drawing/detail/schedule/bills of quantities/specification [*delete as appropriate and add number of drawing, page and item number of bills of quantities etc.*] as follows:

[*Insert details of the divergence.*]

Please let us have your instruction by [*insert date*]. Failure to issue your instruction by that date will result in delay and disruption to the Works for which we will seek appropriate financial recompense.

Yours faithfully

信函 21

致工程师，注意到法定要求与其他文件之间存在矛盾（a)

敬启者：

我方发现法定要求与贵方图纸/细节/计划/工程量清单/规范［酌情删除，增加图纸编号、工程量清单的页码和序号等。］之间存在矛盾，矛盾之处如下：

［插入分歧细节。］

请在［插入日期］前将发布指示给我方。若贵方在该日期前未发出指示，将会导致工程延误与中断，我方将就此要求适当的经济补偿。

敬上

Letter 22

To engineer, noting divergence between statutory requirements and other documents (b)

Dear

We have found what appears to be a divergence between statutory requirements and the Employer's Requirements/our Contractor's Proposals [*delete as appropriate*] as follows:

[*Insert details of the divergence.*]

We propose the following amendment to remove the divergence:

[*Describe in detail how to deal with the divergence including the provision of any drawing.*]

Please let us have your written consent to our proposal by [*insert date*] in order to avoid delay to the Works. Please note the amendment on the contract documents in accordance with clause ____ of the conditions of contract and send us a copy.

Yours faithfully

信函 22

致工程师，注意到法定要求与其他文件之间存在矛盾（b）

敬启者：

我方发现法定要求与“雇主要求”/“承包商建议书”［酌情删除］不一致。矛盾之处如下：

［插入分歧细节。］

为消除不符之处，我方提出如下修改意见：

［请详细描述如何处理以下分歧，包括提供任何图纸。］

为避免工程延误，我方要求在［插入日期］前收到贵方书面同意我方建议。请注意该合同文件的修改应符合合同条款____的要求，并请发给我方一份副本。

敬上

Letter 23

To engineer, if emergency compliance with statutory requirements required

Dear

We hereby give notice as required by clause ____ of the conditions of contract that we have been obliged to carry out work constituting emergency compliance with statutory requirements. The emergency and the steps we are taking/have taken [*delete as appropriate*] are as follows:

[*Insert details*.]

We confirm that the work carried out and materials supplied rank as a variation and we should be pleased to receive any further instructions necessary.

Please take this letter as a notice of delay under clause ____ and an application for loss and/or expense under clause ____ .

Yours faithfully

信函 23

致工程师，如果需要根据法律规定处理紧急情况

敬启者：

按照合同条款第___项的规定，我方特此通知我方必须执行符合法律要求的，构成紧急情况的施工工作。紧急情况及我方正在采取/已采取的措施［酌情删除］如下：

［插入细节。］

我方确认所执行的工作和提供的材料属于工程变更。我方希望收到进一步指示。

根据条款第____项，请视该函为延期通知，并且根据条款第____项，请视该函为损失和/或费用赔偿申请函。

敬上

Letter 24

To engineer, if a change in statutory requirements after base date

Dear

On [*insert date*] there was a change in statutory requirements which necessitates an alteration or modification to the Contractor's Designed Portion [*give details*].

We are putting the necessary amendment in hand and we will write to you again as soon as we are in a position to value the amendment which is to be treated as a variation instruction under the provisions of clause ____ of the conditions of contract. In our view, this is not an instruction to be dealt with under supplemental provision ____ .

[*Then add*:]

Therefore, please take this letter as notice of delay under clause ____ and application for loss and/or expense under clause ____ .

Yours faithfully

信函 24

致工程师，如果法定要求在基准日期后更改

敬启者：

在［插入日期］，法定要求有所更改，因此必须更改或修改承包商设计部分［提供详情］。

我方正在进行必要的修改，一旦我方能够对该修改定价，我方将再次致函贵方。根据合同条款第____项规定，该修改应被视为变更指令。我方认为，这并不是根据补充条款____处理的指示。

［然后补充：］

根据条款第____项，请视该函为延期通知，并且根据条款第____项，请视此为损失和/或费用赔偿申请函。

敬上

Letter 25
To engineer, if development control decision after base date

Dear

We have just received a permission/approval [*delete as appropriate*] from [*state the relevant authority*, *e. g. local planning authority*] and a copy is enclosed. To make them conform, it will be necessary to amend our Contractor's Proposals as follows: [*give details*].

We are putting the necessary amendment in hand and we will write to you again as soon as we are in a position to value the amendment which is to be treated as a change instruction under the provisions of clause ____ of the conditions of contract. In our view, this is not an instruction to be dealt with under supplemental provision ____. Therefore, please take this letter as notice of delay under clause ____ and application for loss and/or expense under clause ____.

Yours faithfully

信函 25
致工程师，如果在基准日期之后颁布开发管理决定

敬启者：

我方刚刚收到［说明有关当局，例如地方规划机构］的许可/批准［酌情删除］，并附上了一份副本。为了使其符合该政策，我方有必要修改“承包商建议书”，具体修改如下：［提供相关细节］。

我方正在进行必要的修改，一旦我方能够对该修改做出定价，我方将再次致函贵方，根据合同条款第____项规定，该修改应被视为变更指令。我方认为，根据附加条款____，该指令并非需要由我方处理。因此，根据条款第____项，请视该函为延期通知，并且根据条款第____项，请视此为损失和/或费用赔偿申请函。

敬上

Letter 26

To employer (not being a local authority), objecting to the nomination of a replacement engineer

Dear

Under the provisions of clause ____ we hereby formally give notice of our objection to the nomination of [*insert name*] of [*insert address*] as engineer for the purpose of this contract in succession to [*insert name and address of previous engineer/PM*].

The grounds for our objections are [*insert particular reasons for objection*].

A good working relationship between engineer and contractor is vital to the successful completion of any project. With this in mind, we look forward to hearing that you have reconsidered the nomination.

Yours faithfully

信函 26

致雇主（不是地方当局），反对提名替换工程师

敬启者：

根据合同条款第____项，我方特此正式通知贵方，我方反对提名［插入地址］的［插入姓名］接替［插入先前工程师/项目经理的姓名与地址］作为本合同的工程师。

我方反对的理由为［插入反对的具体理由］。

工程师与承包商之间良好的工作关系对于任何项目的顺利完成都至关重要。考虑到这一点，我方期待贵方已重新考虑提名。

敬上

Letter 27

To employer (not being a local authority), objecting to the nomination of the employer as replacement engineer

Dear

We were surprised to learn that you proposed to act as engineer yourself as a replacement for the previous engineer. Presumably, you are purporting to act under the provisions of clause ____ . We are not prepared to accept you acting in this role. Not only are you not a person registered with the engineers Registration Board and, therefore, not permitted to use the title 'engineer', we are advised that this kind of provision in a construction contract does not empower you to appoint yourself as engineer and we suggest that you take your own urgent legal advice to confirm that position.

Unless you inform us by close of business on [*insert a date three working days from the date of this letter*] that you have reconsidered the nomination, we shall have no alternative but to seek adjudication on the point and on the damage, loss and expense caused to us by the breach.

Yours faithfully

信函 27

致雇主（非地方当局），反对提名雇主接替工程师

敬启者：

我方惊悉贵方建议亲自接替之前的工程师。贵方大概旨在根据第____条的规定行事。我方不同意贵方担任该工程师一职。贵方并非“工程师注册委员会”注册认定的工程师，因此不允许使用“工程师”头衔。另外，我方获悉施工合同中的该条款未授权贵方任命自己作为工程师。我方建议贵方马上亲自进行法律咨询以确认这一情况。

如贵方未在［插入发送此函三个工作日之后的日期］下班前通知我方贵方已重新考虑提名，我方将只能就此事，以及贵方违约行为给我方造成的损害，损失和费用赔偿事宜寻求裁决。

敬上

Letter 28

To employer, if replacement engineer not appointed

By fax and post

Dear

We were notified/became aware [*delete as appropriate*] that the engineer named in the contract had ceased to act on or about the [*insert date*]. Under the provisions of article ____ you are obliged to nominate a replacement engineer within 21 days.

It is now some [*insert number*] days since you should have made the appointment and you have failed to do so. The result of this is that, at the most basic level, no instructions or certificates can be issued and the extension of time provisions cannot be operated. Therefore, as soon as we require an instruction to enable us to proceed, we shall be obliged to stop that particular activity although we will, for the record, notify you of the instruction required. We envisage that the whole site will be at a standstill very shortly and we must consider our position. In any event, time will soon become at large.

Any replacement engineer will need time to absorb the detail of this project and we urge you to make the appointment immediately. In the meantime you are liable to us for all the damage, loss and expense caused to us by your breach.

Yours faithfully

信函 28

致雇主，如果未任命新的工程师

传真并邮寄

敬启者：

我方获悉/了解到［酌情删除］合同中指定的工程师已于［插入日期］前后离职。根据合同第____条规定，贵方应在21天内提名另一位工程师。

贵方本应在［插入数字］天前任命工程师，但贵方一直未任命。此事至少会造成无法发布指示和证书，也无法操作延长工期的条款。因此，一旦我方需要指示以便能继续施工，我方将无法获得该指示，但为了记录备案，我方将通知贵方我方所需的指示。我方预计整个工地将很快陷入完全停滞状态，我方必须考虑我方的处境。无论如何，工期将非常紧迫。

任何接替的工程师都需要时间了解项目详情，我方敦促贵方立即委派工程师。同时贵方应负责赔偿贵方违约行为给我方造成的损害，损失和费用。

敬上

Letter 29

To employer, objecting to the nomination of a replacement quantity surveyor

Dear

Under the provisions of clause ____ we hereby formally give notice of our objection to the nomination of [*insert name*] of [*insert address*] as quantity surveyor for the purpose of this contract in succession to [*insert name and address of previous quantity surveyor*].

The grounds for our objection are [*insert particular reasons for objection*].

A good working relationship between quantity surveyor and contractor is vital to the successful completion of any project. With this in mind, we look forward to hearing that you have reconsidered the nomination.

Yours faithfully

信函 29

致雇主，反对提名更换估算师

敬启者：

根据合同条款第____项，我方特此正式通知贵方，我方反对提名［插入地址］的［插入姓名］接替［插入先前估算师的姓名与地址］作为本合同的估算师。

我方反对的理由为［插入反对的具体理由］。

估算师与承包商之间良好的工作关系对于任何项目的顺利完成都至关重要。考虑到这一点，我方期待贵方已重新考虑提名。

敬上

Letter 30

To engineer, regarding directions issued on site by the clerk of works

Dear

The clerk of works has issued direction number [*insert number*] dated [*insert date*] on site, a copy of which is enclosed.

Such directions have, of course, no contractual effect. Clearly, the directions of the clerk of works issued in relation to the correction of defective work can be very helpful. We are anxious to avoid misunderstandings on site and in this spirit we suggest that the clerk of works should issue no further directions, other than those relating to defective work. All other matters can be referred directly to you by telephone and, at your discretion, a proper engineer's instruction can be issued.

In our view, the above system would remove a good deal of the uncertainty which must result from the present state of affairs. We look forward to hearing your comments.

Yours faithfully

信函 30

致工程师，关于工程管理员（现场监工）在现场发出的指示

敬启者：

工程管理员［插入日期］在现场发布编号为［插入编号］的指示，我方随函附上一份副本。

此类指示显然无合同效力。当然，工程管理员发布的关于改正施工缺陷的指示可能是有帮助的。我方希望在施工现场避免此类误会。本着这一精神，我方建议，除了与施工缺陷有关的指示之外，工程管理员不应发布任何其他指示。我方可以直接向贵方电话汇报所有其他事项，根据贵方的判断，贵方可以发出适当的工程师指示。

我方认为，上述制度将消除目前情况必然造成的很多不确定性。我方期待听到贵方的意见。

敬上

Letter 31

To engineer, regarding instructions issued on site by the clerk of works

Dear

The clerk of works has issued instruction number [*insert number*] dated [*insert date*] on site, a copy of which is enclosed.

Clause ____ of the conditions of contract provides that the clerk of works may issue only instructions under clause ____ unless you expressly delegate other powers to the clerk of works in writing and notify us under the provisions of clause ____.

The enclosed instruction does not fall under the provisions of clause ____ and we have not yet received a clause ____ notice from you. Our agent on site has strict instructions from us that instructions of the clerk of works are not to be executed unless expressly empowered by the contract. If you wish the instruction to be carried out, please let us have your confirmation without delay or, alternatively, a notice under clause ____ that the clerk of works has power to issue any instructions which the contract empowers you to issue.

Yours faithfully

信函 31

致工程师，关于工程管理员（现场监工）在现场发布的指示

敬启者：

工程管理员［插入日期］在现场发布编号为［插入编号］的指示。我方随函附上一份副本。

合同第____条规定工程管理员只可根据合同第____条要求发布指示，除非贵方以书面形式授予工程管理员其他权力，并依据合同第____条的规定通知我方。

随函附上的指示并不属于条款____中规定事项，并且我方没有收到贵方的条款____通知。我方严格要求现场代理人不得执行工程管理员的指示，除非合同明确授权。如贵方希望我方执行工程管理员指示，请及时向我方提供确认书，或者，按照条款____规定，通知我方工程管理员有权发布任何合同授权贵方发布的指示。

敬上

Letter 32

To engineer, if clerk of works defaces work or materials

Dear

It is common practice for the clerk of works to deface work or materials which are considered to be defective. We assume that the basis for such action is to bring the defect to the notice of the contractor and ensure that it cannot remain without attention.

We object to the practice on the following grounds:

1. The work or materials so marked may not be defective and we will be involved in extra work and the employer in extra costs in such circumstances, because otherwise satisfactory work or materials will have been spoilt.

2. The work or materials so marked, if indeed defective, will not be paid for and will be our property when removed. We may be able to incorporate it in other projects where a different standard is required. Defacement by the clerk of works would prevent such re-use.

We will take no point about the defacing marks we noted on site today, but if the practice continues, we will seek financial reimbursement on every occasion.

Yours faithfully

信函 32

致工程师，如果工程管理员污损工程或材料外观

敬启者：

工程管理员在认为有缺陷的工程或材料上涂画是很正常的。我方认为其目的是为了提醒承包商注意到该缺陷，并确保承包商不要忽视。

我方反对该做法，具体原因如下：

1. 如此标记的工程或材料可能没有缺陷，在此情况下，将给我方造成额外的工作，给雇主造成额外的费用，因为这会破坏原本合格的工程和材料。

2. 如此标记的产品或材料，如果确有问题，我方将得不到付款，并在拆除时其归我方所有。我方还可以将其用于需要不同标准的其他项目中。工程管理员造成的污损将使其无法再利用。

我方不会追究今天项目现场的污损标记行为，但如类似行为继续发生，我方每次都将要求经济补偿。

敬上

Letter 33

To engineer, if numerous 'specialist' clerks of works visiting site

Dear

Clause ____ of the conditions of contract permits the employer to appoint a clerk of works. Your letter of the [*insert date*] informed us that the clerk of works would be [*insert name*]. During the past two weeks a number of persons have presented themselves at our site office purporting to be 'specialist clerks of works' [*or substitute the actual title*] and demanding access to the Works.

These persons are neither known to us nor included in the contract and, therefore, we have exercised our right to exclude them from the Works.

We already permit inspections by those consultants introduced at the project start meeting of the [*insert date*] even though the contract is silent as to their existence, because we desire to extend all reasonable co-operation. To allow 'specialist clerks of works' free access would not, in our view, be reasonable. If you require us to grant them such access, please so inform us in writing, but note that, in such circumstances, we consider that our progress would be hindered and we will take appropriate legal advice in regard to our rights and remedies.

Yours faithfully

信函 33

致工程师，如果很多"专家级"工程管理员参观项目现场

敬启者：

合同条款第____项允许雇主委派一名工程管理员。贵方于［插入日期］来函通知我方所委派的现场监工是［插入姓名］。最近两周，很多人员来到我方施工现场办公室，自称是"专业"工程管理员［或者替代为实际头衔］，并要求进入施工现场。

我方既不认识上述人员，他们也不在合同名单中。因此，我方行使权力，拒绝其进入项目施工现场。

我方已允许出席［插入日期］项目启动会议的有关顾问视察现场，尽管合同并未提及他们的存在，原因是我方希望扩大彼此的合作。我方认为允许"专业工程管理员"自由进出施工现场是不合理的。如果贵方要求我方允许，请以书面形式通知我方。但请注意，我方认为此情况将阻碍施工进程，我方将就我方权利及补救措施咨询律师意见。

敬上

Letter 34

To engineer, if clerk of works instructs operatives direct

Dear

May we draw your attention to an unfortunate situation which is developing on site? The clerk of works, with whom we have had the most cordial relations, is getting into the habit of giving oral directions to our operatives on site even going so far as to reprimand some of them for poor workmanship.

We very much value all the comments of the clerk of works, provided that they are addressed to the person-in-charge/site manager [*delete as appropriate*]. The present situation is causing, as one might expect, a degree of disruption as we have to waste valuable time smoothing ruffled feathers. We have spoken to the clerk of works informally, but to no effect. We are anxious to avoid tension on the site, which would be in no one's best interests and we should be grateful if you would deal with this matter tactfully as soon as possible.

Yours faithfully

信函 34

致工程师，如果工程管理员（现场监工）直接指示施工作业人员

敬启者：

我方提请贵方注意项目施工现场正在出现一些不利情况，那就是与我方关系一直很融洽的工程管理员习惯于对我方现场作业人员口头发号施令，甚至训斥他们施工质量低劣(手艺差)。

如果工程管理员直接向负责人/项目经理［酌情删除］反映意见，我方都十分重视。正如所料，目前的状况对项目施工造成了一定程度的影响，我方不得不浪费宝贵的时间来平复工人的情绪。我方在非正式场合与工程管理员谈过此事，但并无效果。我方迫切希望避免施工现场中对双方都不利的紧张局面。如果贵方能够尽快处理此问题，我方将非常感激。

敬上

第六章　工 期 延 误 信 函

第一节　工 期 延 误 概 述

一、工期延误的含义

工期延误（delay to completion）一般是指对承包商计划完成工程日期的延误，或指对合同规定竣工日期的延误。按照延误的责任划分，延误通常可分为可原谅的延误和不可原谅的延误。可原谅的延误（ excusable delay）是指由于雇主或其雇主人员的行为疏忽所导致的承包商可被原谅的延误。例如，工程师未能及时向承包商提供设计图纸、雇主未能按照合同规定的时间提供现场占有权等导致的延误。不可原谅的延误（ non-excusable delay）是指由于承包商自身行为或不作为所导致的延误。例如，承包商未能提供使工程按期完工的足够的施工人员、承包商未能提供足够的施工设备、承包商因自身的原因拖延工期等。可原谅的延误又可以分为“可补偿的延误（ compensable delay）”和“不可补偿的延误（ non-compensable delay）”，前者是指受到延误的一方，通常为承包商，以雇主或其代理的行为或疏忽为由，有权要求获得金钱补偿的延误；后者是指因中立事件（例如极端恶劣的气候条件）、第三方等导致的不能获得金钱补偿的延误。FIDIC“新红皮书”、“新黄皮书”和“银皮书”，以及美国 AIA 合同、英国 NEC 合同、JCT 合同等都规定了发生工期延误的各种解决办法，这也是解决工期延误的合同基础和依据。承包商如果想在工期延长方面获得有利地位，就需要认真研究与工期延长有关的合同约定，特别是通知期限等程序性规定，积极沟通，以减少工期延误带来的损失。

二、工期延误的原因与种类

工期延误的主要原因有工程变更、不可预见的事件、设计延误、投标程序、雇主原因、资源环境变化等。

（一）工程变更

FIDIC 合同一般规定雇主拥有工程变更的权利，雇主通过委托工程师来行使这项权利。FIDIC“新红皮书”第 13 条赋予了工程师签发变更工程指示的权力，如果承包商认为变更可能延误竣工时间，他应当根据第 20.1 款在 28 天内提出索赔。如果工程数量发生实际变化，且根据 FIDIC 第 12 条可以计量，承包商也可就工期延长提出索赔。

（二）不可预见事件

FIDIC“新红皮书”第 4.12 款规定了不可预见的物质条件，由于建筑和土木工程项目易受外界条件的影响和干扰，因此，不可预见的物质条件或事件就成为施工延误的一个主要原因。根据 FIDIC 合同的定义，不可预见的物质条件的范围十分广泛，包括：

（1）自然物质条件，包括地下和水文条件；

（2）人为的物质障碍，包括罢工、雇主行为等；

（3）其他的物质障碍和污染物，如不可抗力事件等。

承包商在施工过程中遇到不可预见的物质条件时，应及时通知工程师，并遵守该款的其他有关规定，同时承包商有权根据该款规定索赔工期和费用。

（三）设计延误和投标程序

根据 FIDIC“新红皮书”第 1.9 款规定，如果工程师延迟签发施工图纸和指示，承包商有权索赔工期和费用。

（四）雇主原因

如果雇主造成了延误，作为合同另一方的承包商有权要求补偿。FIDIC“新红皮书”第 8.4 款赋予了承包商就工期延长提出索赔的所有权利，并且第 17.3 款和第 17.4 款的雇主风险条款也规定了承包商索赔的权利。

（五）人员或货物的短缺

根据 FIDIC“新红皮书”第 1.1.6.8 款规定，短缺是指一个有经验的承包商无法预见的，并且是由于流行病或政府行为造成的。在出现人员或货物的短缺时，承包商应及时通知工程师，提出人员或货物短缺的证据，并采取相应的措施克服短缺现象。

（六）环境原因

环境原因造成延误主要表现在施工期间不断地发生塌方、流砂，以及连续降雨、洪水暴发、严寒酷暑等环境因素引起的延误。

（七）其他原因

根据 FIDIC“新红皮书”的规定，承包商可以索赔工期的其他原因包括：第 2.1 款现场进入权、第 4.7 款放线、第 4. 24 款化石、第 7.4 款检验、第 10.3 款对竣工检验的干扰、第 13.7 款因法律改变的调整、第 16.1 款承包商暂停工作的权利、第 17.4 款雇主风险的后果以及第 19.4 款不可抗力的后果。

三、承包商工期误期损害赔偿

FIDIC“新红皮书”第 8.7 款规定：“如承包商未能遵守第 8.2 款［竣工时间］的规定，承包商应为其违约行为依照第 2.5 款［雇主的索赔］的要求，向雇主支付误期损害赔偿费。

在 FIDIC 合同中，误期损害赔偿费应具有如下意义：

（1）承包商未能在合同规定的工期内完成工程，则雇主有权扣除投标附录中写明的误期损害赔偿费。如分包商未能在分包合同规定的工期内完成分包工程，则承包商有权扣除分包商报价书附录中写明的误期损害赔偿费。

（2）雇主可以从应付给承包商的款项中扣除误期损害赔偿费，承包商也可从应付给分包商的款项中扣除误期损害赔偿费。

（3）误期损害赔偿费的支付或扣除不应解除承包商完成工程的义务，也不应解除合同规定的任何其他责任和义务。

第二节 工 期 延 误 信 函

Letter 1

To engineer, if delay occurs, but no grounds for extension of time

Dear

The progress of the Works is being/is likely to be [*delete as appropriate*] delayed due to [*state reasons*].

We will continue to use our best endeavours to minimise the delay and its effects and we will inform you immediately the cause of the delay has ceased to operate.

This notice is issued in accordance with clause ____.

Yours faithfully

信函 1

致工程师，如出现延期情况，但没有延长时间的理由

敬启者：

工程进度现正/有可能［酌情删除］面临延期，原因是［说明原因］。

我方将继续竭力缩短延期时间，减少延误所造成的影响。造成延期的问题一旦解决，我方将即刻通知贵方。

依据条款____规定，我方发布该通知。

敬上

Letter 2

To engineer, when cause of delay ended if no grounds for extension of time

Dear

We refer to our letter of the [*insert date*] notifying you of delay.

We are pleased to be able to inform you that the cause of the delay has been dealt with. The measures which we adopted during the period of delay and the continuing procedures over the next few weeks are designed to enable us to recover the lost time and put the progress of the Works on programme by [*insert date*] .

Yours faithfully

信函 2

致工程师，如果没有延期理由，造成延期的问题已解决时

敬启者：

我方于［插入日期］发函通知贵方工期延迟。

很高兴通知贵方造成延期的问题已解决。延期时间内我方所采取的措施和接下来几周内将继续进行的补救工作能使我方弥补损失的时间，在［插入日期］前赶上工程进度。

敬上

Letter 3

To engineer, if delay occurs giving grounds for extension of time (a)

Dear

The progress of the Works is being delayed due to [*state reasons*]. We consider this to be a relevant event under clause [*insert number*].

The delay began on [*insert date*]. When it is finished, we will furnish you with our estimate of delay in the completion of the Works and further supporting particulars.

You may be assured that we are using our best endeavours to prevent delay in progress and completion of the Works.

This notice is issued in accordance with clause ____.

Yours faithfully

信函 3

致工程师，如工程延期，说明延长工期的原因（a）

敬启者：

工程进度由于［陈述理由］而延误，我方认为此事属于合同条款［插入编号］下的相关事项。

延期始于［插入日期］。延期结束时，我方将向贵方提供竣工延误的估计情况，并提供进一步具体细节。

请贵方放心，我方正竭力避免工程进度及竣工延误。

依据条款____规定，我方发布该通知。

敬上

Letter 4

To engineer, if delay occurs giving grounds for extension of time (b)

Dear

It is apparent that the Works will not be complete by the date for completion.

[*Then either*:]

The circumstances are [*state*]. We estimate the delay to total [*insert period*]. We consider that we are entitled to an extension of [*insert period*] and we will furnish further particulars within the next few days.

[*Or*:]

The circumstances are [*state*]. We are unable to estimate the total delay at this stage because it is continuing. When the delay is over, we will submit the required estimate and other details immediately.

[*Then*:]

You may be assured that we are using our best endeavours to prevent delay in progress and completion of the Works.

This notice is issued in accordance with clause ____ of the conditions of contract.

Yours faithfully

信函 4

致工程师，如出现延期，说明延长工期的原因（b）

敬启者：

显然，工程不能如期竣工。

[然后：]

情况为［描述］。据估算，延期总计为［插入时间段］。我方认为我方有权延长［插入时段］，并且会在未来几天内提供进一步的具体信息。

[或者：]

情况为［描述］。我方无法估计现阶段的总延误时间，原因是延期还在持续。问题一经解决，我方将立即呈送所需估算时间和其他细节。

[然后：]

请贵方放心我方正竭力避免工程进度及竣工日期延误。

依据合同条款____规定，我方发布该通知。

敬上

Letter 5

To engineer, providing further particulars for extension of time (a)

Dear

We refer to our letter of the [*insert date*] in which we notified you of a delay in progress of the Works likely to result in a delay to completion of the Works. We note below particulars of the expected effects and the estimated extent of delay in completion of the Works in respect of each relevant event specified in our notice:

[*List relevant events separately, giving an assessment of the delay to completion in each case together with any other supporting information.*]

We believe that you now have sufficient information to enable you to grant a fair and reasonable extension of time.

Yours faithfully

信函 5

致工程师，提供延期的进一步详情（a）

敬启者：

我方于［插入日期］发函通知贵方工程进度延期可能会导致竣工延误。就通知中明确的每个相关事项，我方详细说明竣工延误的预期影响及估计延误的程度，具体如下：

［分别列出相关事项，评估每种情况下竣工延误的情况并提供全部证明信息。］

我方相信贵方目前已充分了解以上信息，能够准许我方合理地延长工期。

敬上

Letter 6
To engineer，providing further particulars for extension of time (b)

Dear

We refer to our notice of the [*insert date*] in which we informed you of a delay in progress of the Works likely to result in a delay to completion of the Works.

We estimate that completion will be delayed by [*insert number*] weeks and we consider that we should be granted an extension of time for that period. We arrive at this conclusion as follows：

[*State reasons in full and include all supporting information.*]

We believe that you now have sufficient information to enable you to grant a fair and reasonable extension of time and we look forward to hearing from you shortly.

Yours faithfully

信函 6
致工程师，提供延期的进一步详情（b）

敬启者：

我方于［插入日期］发函通知贵方工程进度延期可能会导致竣工延期。

我方预计竣工时间会延迟［插入数字］周，望贵方予以批准。造成延期的原因总结如下：

［陈述充足的理由，包括全部证明信息。］

我方相信贵方目前已充分了解以上信息，能够准许我方合理地延长工期。早复为盼。

敬上

Letter 7

To engineer, if requesting further information in order to grant extension of time

Dear

Thank you for your letter of the [*insert date*] in which you request further information in respect of [*state what engineer requires*].

[*State the information required by the engineer in the form of precise answers to the questions or, if the engineer does not say what is required, write as follows*:]

We believe that we gave you all the information you require in our letter of the [*insert date*]. It is not clear from your letter what further information you now request. If you would be good enough to ask specific questions, we will do our best to answer them and supply whatever further supporting details then suggest themselves to us.

We look forward to hearing from you as soon as possible.

Yours faithfully

信函 7

致工程师，如果为了准予延期而要求提供更多信息

敬启者：

感谢贵方［插入日期］来函，贵方要求就［陈述工程师的要求］提供更多信息。

［说明工程师要求的信息内容，准确回答问题；或者，如工程师没有给出要求，书写如下：］

我方已在［插入日期］的函中按贵方要求提供了全部信息。现贵方来函，需要更多信息，但该函并未明确贵方需要哪些信息。贵方最好能提出具体问题，我方将尽力予以解答并提供任何我方能够想到的详细证据。

早复为盼。

敬上

Letter 8

To engineer, if unreasonably requesting further information in order to grant an extension of time

Dear

Thank you for your letter of the [*insert date*] requesting further information in order to enable you to grant an extension of time.

We submitted notice of delay, as required by the contract, on [*insert date*]. We submitted full particulars including estimate of the effect of the delay on completion date on [*insert date*]. We believe that you had all the information necessary to enable you to make a fair and reasonable extension of time by [*insert date*]. It is, of course, very much in our interests to supply you with full information as quickly as possible; this we have done.

It is our view that your latest request for information is nothing but an attempt to postpone the granting of an extension. We, therefore, formally call upon you to carry out your duty under clause ____.

Yours faithfully

信函 8

致工程师，如果无理要求获得更多信息以便准予延期

敬启者：

感谢贵方［插入日期］来函，函中要求我方提供进一步信息，以便贵方可以准许我方延长工期。

依据合同要求，我方于［插入日期］提交了延期通知。我方提交了详细资料，包括延误对竣工日［插入日期］的预计影响。我方相信贵方目前已充分了解以上信息，能够于［插入日期］前准许我方合理地延长工期。当然，我方非常感兴趣的是尽快为贵方提供全部信息；我方已经做到这一点。

我方认为，贵方最近要求获得信息无非是试图推迟准予延期。因此，我方正式要求贵方履行合同____条款下的义务。

敬上

Letter 9

To engineer, if extension of time is insufficient

Dear

We have received today your notification of an extension of time of [*insert period*] producing a new date for completion of [*insert date*].

We find your conclusions inexplicable in the light of the facts and the information we submitted in support of those facts.

Perhaps you would be good enough to reconsider your grant of extension of time or let us have an indication of your reasons for arriving at the time period you have granted.

Yours faithfully

信函 9

致工程师，若延期时间不够

敬启者：

今收到贵方延长工期［插入期限］的通知，新的竣工日期为［插入日期］。

鉴于我方提交的用以支持这些事实的情况与信息，我方认为贵方的决定令人费解。

希望贵方重新考虑准许延期，或让我方了解贵方得出准许延期期限的原因。

敬上

Letter 10

To engineer, who is not willing to reconsider an insufficient extension of time

Dear

Thank you for your letter of the [*insert date*] from which we note that you are not willing to reconsider your grant of extension of time in response to our notice of the [*insert date*] and submissions of further information of the [*insert date*] in response to your request of the [*insert date*].

[*If the engineer has given reasons*:]

We have carefully examined the reasons you give in support of your decision and they reveal that you have ignored much of our submission and the facts of the matter.

[*If the engineer has not given reasons*:]

We note that you refuse to give any reasons for your decision and we can only assume that you are on uncertain ground.

[*Then add*:]

We will be happy to meet you if you think that a full discussion would be helpful. That appears to be the sensible way forward, but failing that, we intend to refer this dispute to immediate adjudication.

Yours faithfully

信函 10

致工程师，不愿重新考虑工程延期时间不够问题

敬启者：

感谢贵方［插入日期］来函，对于我方［插入日期］的通知以及应贵方［插入日期］请求，我方于［插入日期］提交的进一步信息给予了答复。我方注意到贵方不愿重新考虑延长工期许可问题。

［如果工程师给出理由：］

我方仔细研究了贵方做出该决定的理由，发现贵方忽视了我方提交的很多信息及该事项的实际情况。

［如果工程师没有给出理由：］

贵方拒绝就所做决定给出解释，对此，我方只能视之为无正当理由。

［然后补充：］

如果贵方认为充分讨论为有用之举，我方将很高兴与贵方举行会谈。这是当前明智之举，如果不这样做，我方打算立即把此争议送交裁决。

敬上

Letter 11

To engineer, if extension of time not properly attributed

Dear

Thank you for your notification of revision to the completion date.

Unfortunately, you failed to state the extension of time you have attributed to each relevant event. This is now mandatory under clause ____ .

Obviously, having formed your opinion, these details will be readily to hand and we look forward to receiving a revised notification containing this information within the next couple of days.

Yours faithfully

信函 11

致工程师，如果没有对延长工期进行合理归因

敬启者：

很高兴收到贵方关于修改竣工日期的通知。

很遗憾，贵方未能就每一相关事项所导致的延期情况予以说明。根据条款____，此为强制性要求。

显然，在形成贵方意见时，这些详细信息将很容易掌握，我方期望在接下来的几天内收到包含此信息的修订通知。

敬上

Letter 12

To engineer, if extension of time not granted within time stipulated

Dear

Notice of delay was sent to you on [*insert date*] in accordance with clause ____ of the conditions of contract. Full particulars including estimate of the delay to completion and estimate of the extension required were sent to you on [*insert date*]. You made no request for further information. Clause ____ requires you to notify us of your decision not later than 12 weeks from receipt of particulars. The period elapsed on [*insert date*] and you have not informed us of your decision. You are clearly in breach of contract, a breach for which the employer is responsible.

You are not empowered to make any extension of time for the current relevant events until after the contract completion date. Any loss or expense which we may suffer, whether from increasing resources or otherwise, as a result of your breach will be recovered from the employer as damages in due course.

Yours faithfully

Copy: Employer

信函 12

致工程师，如未能在规定时间内准予延长工期

敬启者：

根据合同条款____，延期通知已于［插入日期］寄送至贵方。估算的延期竣工时间和所需延期时间的全部详情已于［插入日期］发出。贵方并未要求进一步的信息。根据条款____，贵方需在收到详情起 12 周内通知我方贵公司的决定。［插入日期］已过，贵方并未将贵方决定告知我方，这已构成违约，雇主需对此负责。

在合同完成日期之前，贵方无权对当前相关事项进行任何延长时间。贵方违约会造成我方资源或其他方面的浪费，雇主应及时对我方的损失和额外费用做出相应赔偿。

敬上

抄送至：雇主

Letter 13

To engineer, if slow in granting extension of time

Dear

Notice of delay was sent to you on [*insert date*] in accordance with clause ____ of the conditions of contract. Full particulars were sent to you on [*insert date*].

You have now had [*insert number*] weeks in which to make your decision and we now call upon you to grant us the extension of time to which we are entitled. If we do not receive your notice granting such extension by [*insert date*] you will be in breach of contract, a breach for which the employer will be responsible.

[*If using IC or ICD, add:*]

You will not be empowered to make any extension of time for the current events until after the date of practical completion. Any loss or expense which we may suffer, whether from increasing resources or otherwise, as a result of your breach will be recovered from the employer as damages in due course.

[*If using MW* 98, *add:*]

We will consider that time is 'at large' and the employer will have lost the right to deduct liquidated damages because there will be no date for completion from which liquidated damages can run and you will have lost your power to fix such a date. Our obligation will then be to finish the Works within a reasonable time.

Yours faithfully

Copy: Employer

信函 13

致工程师，如未及时准予工期延长

敬启者：

根据合同条款____，延期通知已于［插入日期］寄送贵方。全部详情已于［插入日期］发出。

贵方现在已有［插入数字］周可以做出决定，现要求贵方准许我方延长工期，我方有权延长工期。若我方未能于［插入日期］前收到准予延期通知，贵方将违约，雇主将对此违约负责。

［若使用 IC 或 ICD，请补充：］

在实际竣工日期之前，贵方将无权对目前相关事项延长工期。贵方违约会造成我方遭受资源增加及其他方面的损失，该损失将在适当时候从雇主处追索，作为我方的损害赔偿。

［若使用 MW 98，请添加：］

我方认为贵方工期为“大约时间”，雇主将失去扣除违约赔偿金的权利，原因是将没有扣除违约金的起始竣工日期，贵方将失去确定竣工日期的权利。我方义务是在合理时间内完成工程。

敬上

抄送至：雇主

Letter 14

To engineer, if review of extensions not carried out (a)

Dear

Clause ____ requires you to either:

1. Fix a completion date later than that previously fixed, or

2. Fix a completion date earlier than that previously fixed, or

3. Confirm the completion date previously fixed.

You must carry out this duty, at latest, within 12 weeks after the date of practical completion. We are advised that, despite speculation to the contrary, the time period is mandatory. That period expired on [*insert date*]. You are in breach of your duty and our obligation now is simply to complete within a reasonable time. The employer has lost the right to deduct liquidated damages, because there is no date fixed for completion from which such damages can run and you have lost your power to fix such a date. Any attempt to deduct such damages will result in immediate legal action on our part.

Yours faithfully

Copy: Employer

信函 14

致工程师，若没有执行延期审核（a）

敬启者：

条款____要求贵方做到以下其中一点：

1. 确定一个比预先确定时间稍晚的竣工日期，或；

2. 确定一个比预先确定时间稍早的竣工日期，或；

3. 确认先前确定的竣工日期。

最迟在实际竣工日期后的12周内，贵方必须履行该义务。尽管与预测情况相反，但我方认为工期是强制性的。最后期限［插入日期］已过。贵方没有履行职责，现在我方责任仅为在合理时间内完工。雇主已丧失扣除违约金的权利，因为没有规定竣工日期，也就没有损害赔偿金问题，并且贵方无权确定竣工日期。任何扣除此类损害赔偿的企图将导致我方立即采取法律措施。

敬上

抄送至：雇主

Letter 15

To engineer, if review of extensions not carried out (b)

Dear

Clause ____ permits you to extend time whether upon reviewing previous decisions or otherwise and whether or not we have given notice of delay. This is a valuable power for you to review the situation after the works are finished.

However, you must carry out this duty within 12 weeks after the date of practical completion. We are advised that, despite speculation to the contrary, the time period is mandatory. That period expired on [*insert date*] and our obligation now is simply to complete within a reasonable time. The employer has lost the right to deduct liquidated damages, because there is no date fixed from which such damages can run and you have lost your power to fix such a date. Any attempt to deduct such damages will result in immediate legal action on our part.

Yours faithfully

Copy: Employer

信函 15

致工程师，如没有执行延期审核（b）

敬启者：

无论是否审查以前的决定还是出于其他原因，无论我方是否给出延期通知，条款____允许贵方延长工期。竣工后，贵方有审核该情况的权力。

然而，最迟在实际竣工日期后的12周内，贵方必须履行该义务。尽管与预测情况相反，但我方认为工期是强制性的。最后期限［插入日期］已过。贵方没有履行职责，现在我方责任仅为在合理时间内完工。雇主已丧失扣除违约金的权利，因为没有规定竣工日期，也就没有损害赔偿金问题，并且贵方无权确定竣工日期。任何扣除此类损害赔偿的企图将导致我方立即采取法律措施。

敬上

抄送至：雇主

Letter 16

To engineer, if no final decision on extensions of time

Dear

Clause ____ of the conditions of contract stipulates that you must come to a final decision on all outstanding and interim extensions of time within 42 days after completion of the Works. We are advised that, despite speculation to the contrary, the time period is mandatory. The period expired on [*insert date*] and you have not made a final decision on the request (s) for extension of time originally notified to you on [*insert date or dates*] and our obligation now is simply to complete within a reasonable time. The employer has lost the right to deduct liquidated damages, because there is no date fixed for completion and you have lost your power to fix such a date. Any attempt to deduct such damages will result in immediate legal action on our part.

Yours faithfully

Copy: Employer

信函 16

致工程师，如果没有最后决定延长工期

敬启者：

合同条款____规定贵方必须在竣工后 42 天内对所有未完成的及临时的延期作出最终决定。尽管与预测情况相反，但我方认为工期是强制性的。最后期限［插入日期］已过。贵方尚未就我方最初在［插入日期］提交的延期请求作出最终决定，现在我方责任仅为在合理时间内完工。雇主已丧失扣除违约金的权利，因为没有规定竣工日期，也就没有损害赔偿金问题，并且贵方无权确定竣工日期。任何扣除此类损害赔偿的企图将导致我方立即采取法律措施。

敬上

抄送至：雇主

Letter 17

To engineer, who alleges that contractor is not using best endeavours

Dear

Clause ____ of the conditions of contract requires us to use our best endeavours to prevent delay. Your allegation in [*state where and date, e.g.: minute no. 7.4 of the site meeting held on the* 3 *September* 2007] that we are failing to carry out our duties in this respect is totally without foundation. Our obligation to use best endeavours is simply an obligation to continue to work regularly and diligently, rearranging our labour force as best we can. This we have done and we are continuing so to do. There is no obligation upon us to expend additional sums of money to make up lost time. If that was the case, the extension of time clause would be otiose.

If you purport to reduce, on grounds of failure to use best endeavours, our entitlement to an extension of time, we will take immediate and appropriate advice on the remedies available to us.

Yours faithfully

信函 17

致工程师，工程师称承包商未竭尽全力

敬启者：

合同条款____要求我们双方应竭力避免工程延期。贵方于［说明时间，地点，例如：2007 年 9 月 3 日编号为 7.4 的工地现场会议］称我方未能履行该职责，此指责毫无根据。我方所谓的竭尽全力履行职责是指尽我方所能继续按计划努力施工，及重新分配劳动力，并保证工程质量。我方已经并将继续履行该职责。我方无义务投入额外资金以弥补失去的时间。如当时情况属实，则延期条款应该无效。

如果贵方以我方未能尽最大努力为由而消减我方的延期权利，我方将就我方可用的补救措施立即采取适当法律措施。

敬上

Letter 18

To engineer，if non-completion certificate or notice wrongly issued

Dear

We are in receipt of what purports to be a non-completion certificate issued under clause ____ . The issue of a valid certificate is a pre-condition to the deduction of liquidated damages. In this instance，it is not valid because

[*Either*:]

you have not yet given the correct extension of time.

[*Or*:]

it was issued prior to the completion date.

[*Or*:]

it does not comply with the terms of the contract.

[*Then add* :]

If the employer attempts to withhold liquidated damages，a dispute will then arise which we will refer to adjudication. In order to avoid the resultant expense to both parties，we suggest that you withdraw your certificate immediately.

Yours faithfully

信函 18

致工程师，如果错误发出未竣工证书或通知

敬启者：

收悉贵方依照条款____发出的未竣工证书。签发有效证书是扣除违约金的先决条件。在此情况下，该证书无效，原因如下：

[或者:]

贵方未提供正确的延期时间。

[或者:]

该证书先于竣工日期发出。

[或者:]

该证书与合同条款不符。

[然后补充:]

如雇主试图扣留违约金，则会出现争议，我方将把此争议提交裁决。为避免造成双方费用损失，建议贵方立即撤销证书。

敬上

Letter 19

To employer, if liquidated damages wrongfully deducted

Dear

We have received your cheque today in the sum of [*insert amount*] which falls short of the amount certified as due to us in certificate number [*insert number*] by [*insert amount*]. We note from your accompanying letter of the [*insert date*] that the deficit represents your deduction of alleged liquidated damages in the sum of [*insert amount*] for [*insert number*] weeks.

We consider you to be in breach of contract because [*state reasons*]. If we do not receive your cheque for the full amount of [*insert amount*] by [*insert date*], we will take appropriates to recover not only the amount wrongfully deducted, but also damages, interest and costs. We reserve the right to suspend performance of all obligations under clause ____ and/or to terminate our employment under clause ____.

Yours faithfully

信函 19

致雇主，如果扣除违约金有误

敬启者：

今收到贵方支票，金额为［插入金额］，该金额少于编号为［插入编号］证书中的应付给我方的核证金额［插入数额］。从贵方［插入日期］随函得知，赤字表示贵方扣除我方［插入数字］周的，金额为［插入数额］所谓违约金。

我方认为贵方已构成违约，原因是［陈述理由］。如果我方未能于［插入日期］前收到全额［插入数额］支票，我方将采取适当措施，不仅要追索被错误扣除的金额，还要追索赔偿损失，利息和费用。我方保留中止履行合同____条款下的所有义务的权利，并/或根据条款____解除双方的雇用关系。

敬上

Letter 20

To employer, if damages repaid without interest

Dear

We are in receipt of your letter of the [*insert date*] enclosing your cheque for [*insert amount*] representing liquidated damages wrongfully deducted/recovered [*delete as appropriate*]. You will recall that in our letter dated [*insert date*] we gave you due notice that we should require damages for your breach. We are prepared to accept interest charges as a realistic basis for damages and the additional payment required on account of such interest is, therefore, calculated on the basis of 5% above Bank of England Base Rate current at the date payment became overdue [*insert amount together with the calculation*].

We assume that this is a genuine oversight on your part and, while reserving all our rights and remedies, we do not propose to take any action if we receive the sum of [*insert amount of interest*] by [*insert date*].

Yours faithfully

Copy: Engineer

信函 20

致雇主，如果赔偿损失而没有支付利息

敬启者：

收悉贵方［插入日期］来函及附寄的金额为［插入数额］的支票，此为贵方错误扣除/追缴［酌情删除］的违约金。贵方会记得，我方于［插入日期］正式函告贵方我方要求获得违约赔偿。我方准备接受利息费用作为损害赔偿的实际依据，因此，由该利息所导致的额外付款的计算依据是高于逾期付款日［插入金额及计算方式］英格兰银行基准利率5%。

我方认为此事确为贵方过失所致，虽然我方保留所有权利和补救措施，但如果我方在［插入日期］前收到金额为［插入利息额］的款项，我方不打算采取法律措施处理此事。

敬上

抄送至：工程师

第七章 款项支付信函

第一节 款项支付概述

一、合同价格与支付

合同价格（The Contract Price）是指中标后，雇主与承包商签订的建筑工程合同中体现的雇主应向承包商支付的金额。而支付是雇主根据合同约定金额（包括但不限于工程量清单等付款凭证）向承包商的付款。合同价格是支付的基础，雇主根据合同价格和其他依据对承包商进行支付。对于承包商而言，稳定有保障的支付是其施工运行与管理中最为复杂的，并需要良好技巧加以解决的事情，因此灵活而有效的信函沟通显得十分重要。根据 NEC、FIDIC 等合同文件的规定，如果工程师在规定的期限内没有签发付款证书，这是违背合同的行为，雇主要为此承担责任，承包商因此可以得到雇主的赔付；如果雇主对承包商拒绝支付，承包商也可以因此解除与雇主的合同，雇主需要对承包商进行适当的赔付。在工程实践中，因支付引起的纠纷构成了工程施工合同纠纷的主要部分，有些纠纷可以通过工程师调解，或双方协商得到解决；有些纠纷则会进一步发展，双方可能会选择通过仲裁或司法程序加以解决。这样解决纠纷的道路会变得十分漫长，而由此可能给承包商带来的影响或损失是不可估量的。本章主要侧重临时付款、雇主延迟付款、付款证书、最终付款等信函的写作。

二、FIDIC 合同支付的种类

（一）开工预付款

开工预付款是指工程承包合同签订后，承包商向雇主呈交了已获得认可的履约保证书及保函后，工程师开具支付证明，由雇主向承包单位支付的工程预付款，支付的比例可由承包商与雇主在签订合同时商定，预付款的比例一般不高于合同总价的 12%。在国际工程承包中，承包商一般在项目的启动阶段需要投入大笔的资金，为了帮助解决承包商启动资金的困难，FIDIC 合同规定，雇主应向承包商支付一定数额的预付款，故此时的预付款，又可称为动员预付款。开工预付款支付的具体情况如支付比例、分期支付的次数、支付时间、支付货币及货币比例等，由双方来确定，承包商应提交开工预付款保函。

（二）材料预付款

FIDIC“新红皮书”第 14.5 款规定了“用于永久工程的设备和材料”。在 FIDIC 合同中，为了帮助承包商解决订购大量材料和设备占用资金周转的困难，规定雇主在一定条件下应向承包商支付材料、设备预付款。通用条件中规定一般材料、设备预支额度为实际费用的 80%。

（三）中间支付——BOQ 支付

BOQ 是 Bill of Quantity 的缩写，即工程量清单，是指工程按项目进行量化的支付，一般施工过程中的常规支付都采取这种方式，如开挖量、混凝土量、灌浆等项目。它是主要的中间支付形式。BOQ 项目的分项一般由雇主在招标时以工程量清单的形式提供，承包商在投标时根据施工布置及现场情况等条件报价，经双方认可后的报价写入合同作为施工过程中间支付的依据。值得说明的是，由雇主确定的工程量，是承包商报价的重要基础，除表层开挖、基础处理这些工程量变化很大的项目外，工程量清单中其他项目的工程量不能与实际偏离太大，一般不宜超过25%，否则会引起索赔。BOQ支付一般由承包商在每次的中间支付申请中提出申请，由工程师负责审核发生的实际工程量，按照合同中工程量清单中明确的单价予以支付。

（四）保留金

保留金是按合同约定从承包商应得的工程进度款中相应扣减的一笔金额，它保留在雇主手中，作为约束承包商严格履行合同义务的措施之一。当承包商有一般违约行为使雇主受到损失时，可从该项金额内直接扣除损害赔偿费。如果承包商未能在工程师规定的时间内修复缺陷工程部位，雇主雇用其他人完成后，这笔费用可从保留金内扣除。根据FIDIC"新红皮书"第14.9款的规定，当工程师已经颁发了整个工程的接收证书时，工程师应开具证书将保留金的一半支付给承包商。如果颁发的接收证书只是限于一个区段或工程的一部分，则应就相应百分比的保留金开具证书并给予支付。这个百分比应该是将估算的区段或部分工程的合同价值除以整个工程最终合同价格的估算值计算得出的比例的40%。

（五）竣工结算

FIDIC合同规定承包商收到工程接收证书后84天时间内，提交工程竣工报表。工程师接到竣工报表后，应对照竣工图进行工程量的详细核算，对其他支付要求进行审查，然后再依据审查结果签署竣工结算的支付证书。此项签证工作，工程师也应在收到竣工报表后28天内完成。雇主根据工程师的签证予以支付。

（六）最终结算

最终结算是指颁发履约证书后，对承包商完成全部工作价值的详细结算以及根据合同对应付给承包商的其他费用进行核实，确定合同的最终价格。颁发履约证书后的56天内，承包商应向工程师提交最终报表草案以及工程师要求提交的有关资料。最终报表草案要详细说明根据合同完成的全部工程价值和承包商依据合同认为还应支付给他的任何进一步款项，如剩余的保留金及缺陷通知期内发生的索赔费用等。工程师审核后与承包商协商，对最终报表草案进行适当的补充或修改后形成最终报表。承包商将最终报表送交工程师的同时，还需向雇主提交一份"结清单"，进一步证实最终报表中的支付总额，作为同意与雇主终止合同关系的书面文件。工程师在接到最终报表和结清单附件后的28天内签发最终支付证书，雇主应在收到证书后的56天内支付。

三、AIA合同支付的特点

AIA合同文件对雇主的支付能力作出了明确的规定，AIA合同规定，按照承包商的书面要求，工程正式开工之前，雇主必须向承包商提供一份合理的证明文件，说明雇主方已根据合同开始履行义务，做好了用于该项目资金调配的准备工作。提供这份证明文件是工程开工或继续施工的先决条件。证明文件提供后，如未通知承包商，雇主的资金安排不得再轻易变动。该规定可以对雇主的资金准备工作起到一定的推动和监督作用，同时也说明AIA合同在雇主和承包商的权利义务分配方面处理得比较公正合理。

AIA合同在承包商申请付款的问题上较倾向于承包商。例如，AIA合同规定在承包商没有过错的情况下，如果建筑师在接到承包商付款申请后7日内不签发支付证书，或在收到建筑师签发支付证书情况下，雇主在合同规定的支付日到期 7 日未向承包商付款，则承包商可以在下一个 7 日内书面通知雇主和建筑师，将停止工作直至收到应得的款额，并要求补偿因停工造成的工期和费用损失。与 FIDIC 相比，AIA 合同从承包商催款到停工的时间间隔更短，操作性更强。AIA 在时间上的限定和停工后果的严重性会促使三方避免长时间“扯皮”，特别是雇主，在面临巨大的停工压力时，会迅速解决付款问题，这体现了美国工程界的效率，也是美国建筑市场未造成工程款严重拖欠的原因之一。

第二节 款项支付信函

Letter 1

To engineer, enclosing interim application for payment (a)

Dear

Under the provisions of clause ____, we enclose an application for interim payment stating the amount due to us calculated in accordance with clause ____ . In support of our application we enclose the following details as required under clause ____:

[*List*]

We draw your attention to clause ____ which stipulates that you must pay the amount stated as due in this application within 14 days of its receipt.

Yours faithfully

信函 1

致工程师，随函附上临时付款申请（a）

敬启者：

根据条款____规定，我方附上一份临时付款申请，说明按照第____条款计算出的应付给我方的款项金额。兹附寄条款____要求的下述细目用来支持我方的申请。

[列出详情]

我方提请贵方注意合同条款____，该条款规定贵方必须在收到申请后 14 天内支付本申请中所述的金额。

敬上

Letter 2
To engineer，enclosing interim application for payment（b）

Dear

We enclose an application for progress payment stating the amount due to us calculated in accordance with the provisions of clause ____. In support of our application，we enclose the following documents：

［*List*］

Yours faithfully

信函 2
致工程师，随函附上临时付款申请（b）

敬启者：

随函附寄一份工程施工进度付款____的申请，说明按照第____条款计算出的应付给我方的款项金额。兹附寄条款____要求的下述细目用来支持我方的申请。

［列出详情］

敬上

Letter 3

To quantity surveyor, submitting valuation application

Dear

We enclose an application setting out what we consider to be the valuation calculated in accordance with clause ____ .

May we remind you that, under clause ____ , you must now make a valuation for the purpose of ascertaining the amount due in an interim certificate and, to the extent that you disagree with our application, you must submit to us a statement which identifies such disagreement. Note that the statement must be in similar detail to that contained in our application.

Yours faithfully

信函 3

致估算师，提交估价申请

敬启者：

我方随函附寄一份申请，列明我方认为是根据条款____计算所得的估价。

请贵方注意，按照条款____，贵方现在必须进行估价，以确定临时付款证书中的应付金额，并且，如果贵方不同意我方的申请，必须向我方提交一份确认该分歧的说明。请注意，该说明必须与我方申请书中的情况说明同样详细。

敬上

Letter 4

To engineer, if quantity surveyor fails to respond to the valuation application

Dear

We acknowledge receipt of certificate number [*insert number*] dated [*insert date*] from which we see that the amount valued was substantially less than the amount included in our application submitted under the provisions of clause ____. We are concerned because, contrary to the express provisions of the clause, the quantity surveyor has not submitted to us a statement, in similar detail to our application, identifying any disagreement with our application. This should have been done at the time of the quantity surveyor's valuation, i. e., no later than 7 days before the date of the next certificate.

The quantity surveyor's failure amounts to a breach of contract for which we are entitled to appropriate damages. When we did not receive the quantity surveyor's detailed disagreement by the valuation date, we were entitled to assume that the quantity surveyor did not disagree. Accordingly and in reliance, we undertook expenditure of sums of money which we now realise we are not going to receive.

We suggest that damages could be avoided if you would immediately issue a supplementary certificate for the difference between the two valuations.

Yours faithfully

信函 4

致工程师，如果估算师未对估价申请作出回复

敬启者：

收到贵方［插入日期］的［插入编号］证书，我方注意到估值金额明显少于我方按照条款____所提交的申请中的数额。我方甚感担忧，原因是，估算师违反该条款的明确规定，未能向我方提交情况说明，该说明应与我方申请同样详尽，详细说明任何与我方申请相左的意见。此事应该在估算师估价时完成，即不迟于下一个证书发出日期之前 7 天。

估算师未做到这一点已构成违约，对此我方有权要求适当赔偿。鉴于未能在估值日前收到估算师的详细分歧说明，我方有权假定估算师无异议。因此，出于信任，我方已经支出了费用，我方现在意识到该笔支出将无法收回。

我方认为倘若贵方立即就两个估价之间的差额发出补充证明书，损失是可以避免的。

敬上

Letter 5

To engineer，if interim certificate not issued

Dear

We have not received a copy of the interim certificate which should have been issued on the [*insert date*] in accordance with clause ____ of the conditions of contract.

It may be that the certificate has been lost in the post and on this assumption we should be pleased if you would send us a further copy.

If you have not，in fact，issued a certificate，we must remind you of your contractual duty so to do and request that it is in our hands by [*insert date*]. Failing which，we will take immediate legal action against the employer for the breach.

Yours faithfully

Copy：Employer

信函 5

致工程师，如果没有颁发临时证书

敬启者：

我方仍未收到临时证书副本，依照合同条款____规定，该证书本应于［插入日期］颁发。

该证书可能在邮递时丢失，若如此，请贵方再邮寄一份副本。

如果贵方仍未颁发证书，我方必须提醒贵方履行该合同义务，并要求于［插入日期］前收到该证书。否则，我方将立即针对雇主的违约行为采取法律措施。

敬上

抄送至：雇主

Letter 6

To engineer, if certificate insufficient

Dear

We have received your interim certificate number [*insert number*] dated [*insert date*]. We note [*insert the disputed figures*] which do not correspond with the evidence in the documents we have submitted, our discussions with you/the quantity surveyor [*delete as appropriate*] or the situation on site.

If we do not receive a corrected certificate by [*insert date*] we shall seek immediate adjudication on the matter.

Yours faithfully

Copy: Employer

信函 6

致工程师，如果临时证书不充分

敬启者：

我方收到日期为［插入日期］的［插入编号］临时证书。我方注意到［插入有争议的数据］，该数据与我方所提交文件中的证据不符，与我方和贵方/估算师［酌情删除］的磋商不符，或与工地现场情况不符。

如果我方未能于［插入日期］前收到更正证书，我方将立即就此事寻求裁决。

敬上

抄送至：雇主

Letter 7

To employer, if payment not made in full and no withholding notice issued

Special/recorded delivery

Dear

We have today received your cheque for [*insert amount*], some [*insert difference*] less than certified due to us in the engineer's certificate number [*insert number*] dated [*insert date*]. We note that you have withheld [*insert amount*], but you have given no reasons for doing so/the notice you have sent is out of time/the notice you have sent does not properly particularise the grounds for withholding [*delete as appropriate*].

Therefore, your action amounts to a breach of contract and it is contrary to the provisions of the Housing Grants, Construction and Regeneration Act 1996. If we do not receive the sum of [*insert the amount withheld*] by [*insert date*] we will exercise our contractual or other remedies as we deem appropriate to recover the full amount due.

Yours faithfully

Copy: Engineer

信函 7

致雇主，如未全额付款且未发布扣款通知

快递/挂号邮件

敬启者：

现收到贵方［插入金额］的支票，比工程师［插入日期］编号为［插入号码］的证书认定的应付我方金额少［插入款项差额］。我方注意到贵方扣缴了［插入金额］，但贵方没有说明扣款理由/贵方所发出的通知已过期/贵方所发布的通知没有正确说明扣款缘由［酌情删除］。

因此，贵方行为已构成违约，且违反了《1996 年房屋转让、施工和重建法案》中的规定。如未能于［插入日期］前收到金额为［插入扣缴金额］款项，我方将采取我方认为合适的合同补救措施或其他补救措施收回应付金额。

敬上

抄送至：工程师

Letter 8

To engineer, regarding copyright if payment withheld

Dear

We note that you persist in withholding payment of [*describe the situation*].

Copyright in the design of the contractor's designed portion vests in this company.

Clause ____ expressly provides that copyright in all designs, drawings and other documents which we provide vests in this company.

You will only acquire a licence to reproduce such designs in the form of a building when you have paid the amounts you owe. Until then and for the avoidance of doubt take this as notice that you are currently infringing our copyright and we do not grant you a licence. If you do not pay us by close of business on [*insert date*] the money owing, we shall seek to recover substantial damages from you in respect of the infringement.

Yours faithfully

信函 8

致工程师，付款被扣留情况下的相关版权问题

敬启者：

我方注意到贵方执意拒付［描述情况］款项。

承包商设计部分的设计版权归属于该公司。

条款____明确规定我方所提供的全部设计，图纸及其他文件的版权都归属于该公司。

贵方支付了所欠款额后，将只能获得以建筑物的形式复制该设计的许可。在此之前，为避免疑义，我方特此通知贵方目前一直侵犯我方版权，且我方未授予贵方许可。倘若贵方未能于［插入日期］下班前偿还所欠款额，我方将就侵权问题要求贵方对我方的实质性损害予以赔偿。

敬上

Letter 9

To employer, if the advance payment is not paid on the due date

Dear

Under the provisions of clause ____ an advance payment in the sum of [*insert the amount*] was due to be paid to us on the [*insert date*] as stated in the contract particulars. That is [*insert number*] days ago. We furnished a bond on the [*insert date*] in the standard terms from a surety which you have approved.

Your failure to provide the advance payment is a serious breach of contract. We were relying on the payment to assist our funding of this project. Indeed, the offer of such payment was the deciding factor in deciding to enter into this contract.

We are advised that we are entitled to damages for your breach and that, if it continues for more than a few days, we may be able to treat it asrepudiatory, because it will effectively prevent us from proceeding with the Works. In that case, we would be able to accept the breach and bring our obligations to an end. Hopefully, that will not be necessary and we look forward to receiving your payment by return.

Yours faithfully

信函 9

致雇主，如未能如期支付预付款

敬启者：

根据条款____，预付款［插入金额］理应按合同细则所述于［插入日期］支付给我方，但至今已过［插入数字］天。我方于［插入日期］按照贵方认可的担保人的标准条款要求缴纳了保证金。

贵方未能支付预付款已构成严重违约。我方一直靠此款协助我方为该项目融资。事实上，贵方提供该预付款是我方决定签订本合同的决定因素。

我方知悉我方有权就贵方违约行为要求损害赔偿，如此情况再持续几日，我方有可能视其为毁约，因为这会造成我方无法继续施工。在这种情况下，我方能够接受违约，并不再履行我方义务。我方希望事态不必发展至此，希望尽快收到贵方款项。

敬上

Letter 10

To engineer, if valuation not carried out in accordance with the priced activity schedule

Dear

We have just received certificate number [*insert number*] dated [*insert date*] from which we see that the amount certified is substantially less than we expected. It appears that the valuation has not been carried out using the priced activity schedule we supplied.

May we draw your attention to clause ____ which specifically provides that where there is an activity schedule, the value of work in each activity to which it relates must be a proportion of the price stated for the work in that activity equal to the proportion of the work in that activity that has been properly executed.

A simple breakdown is enclosed showing the figure produced by applying that calculation. No doubt the matter is simply the result of an oversight. However, in view of the serious shortfall in the amount certified, we believe that a further and immediate supplementary certificate should be issued to rectify the position.

Yours faithfully

信函 10

致工程师，若没有依照定价明细表实施估价

敬启者：

我方刚刚收到［插入日期］的编号为［插入数字］的证书，并注意到认证金额明显低于我方预期。很明显，贵方并未使用我方提供的定价明细表进行估价。

请贵方注意条款____，该条款明确规定在有施工安排的情况下，与之相关的每项安排中的施工价格必须是该施工安排报价的一部分，且该价格与该安排中已经顺利完成的施工部分相符。

随函附上简明细目，用以说明据此计算所得的金额。该问题无疑仅为疏忽所致。然而，鉴于认证金额严重不足，我方建议贵方立即补发额外证书来纠正此事。

敬上

Letter 11

To engineer, if contractor not asked to be present at measurement

Dear

Looking at the latest valuation, we are concerned to note that although it obviously concerns work which the quantity surveyor has had to measure for valuation purposes, we were not given the opportunity to be present while the measurement was carried out. This is contrary to the provisions of clause ____. The work to which we refer is [*name or describe the work*].

The work in question is now covered up. We should be pleased to receive copies of your measurement notes. To the extent that we do not agree your measurements, we believe that your breach leads to a presumption that our quantities are correct.

We look forward to hearing from you.

Yours faithfully

Copy: Quantity surveyor

信函 11

致工程师，测量时，如未要求承包商在场

敬启者：

看到最新估价，我方注意到，虽然该价格显然涉及估算师为估价目的必须测量的施工项目，但实施测量时，我方未被邀请在场。此举违反条款____规定。我方所指的工程为［*工程名称或描述工程*］。

相关施工项目已经被覆盖，我方希望收到贵方测量记录副本。就此情况来说，我方不同意贵方的测量结果。我方认为贵方违约，据此推定我方评估的工程量是正确的。

早复为盼

敬上

抄送至：估算师

Letter 12

To engineer, requesting payment for off-site materials

Dear

The following goods and materials are 'Listed Items' and are stored at [*insert place*] and available for your inspection at any time:

[*List*]

We should be pleased if you would operate the provisions of clause ____ of the conditions of contract and include the value of such goods and materials in your next interim valuation. We confirm that we have complied with all the requirements of the contract in respect of such goods and materials which the enclosed documents prove are our property.

[*Add, if appropriate:*]

We confirm that we have provided a bond in terms as annexed to the contract.

Yours faithfully

信函 12

致工程师，请求支付非现场材料费

敬启者：

以下物品和材料皆为“开列项目”，储存于［添加地点］，贵方可随时检查。

［列出清单］

如贵方能执行合同条款____规定，并在下次临时估价中包括这些物品和材料的价格，我方将甚感欣然。我方确认我方已遵守合同中关于该货品及材料的所有规定，所附文件证明其为我方财产。

［若合适，请补充：］

我方确认已按照合同附件中的条款要求提供了担保。

敬上

Letter 13

To employer，giving 7 days notice of suspension

Special delivery

Dear

We note that you have not issued any withholding notice，but you have failed to pay/pay in full [*delete as appropriate*] the sum due by the final date for payment. Without prejudice to our other rights and remedies，we give notice that unless you pay the amount due in full within 7 days after receipt of this letter，we shall，in accordance with clause ____，suspend performance of all our obligations under the contract until full payment is received.

Yours faithfully

Copy：Engineer

信函 13

致雇主，发出停工 7 天通知

快递

敬启者：

我方注意到贵方未发布任何扣款通知，但在最后付款日期前没能支付/全额支付［酌情删除］应付工程款。在不影响我方其他权利和补救措施的前提下，我方通知贵方如贵方在收到该函起 7 日内未全额支付应付款项，我方将暂停履行合同规定的所有义务，直到收到全额付款为止。

敬上

抄送至：工程师

Letter 14

To employer, if payment in full has not been made within 7 days despite notice of suspension

Dear

Further to our letter dated [*insert date*] stating that if payment in full was not made within 7 days from the date of its receipt we would suspend our obligations, we have not received any/full [*delete as appropriate*] payment. This is to inform you that, with immediate effect, we are suspending all our obligations under the contract.

We have left the site safe, but it is now for you to arrange security and insurance if you persist in withholding payment.

Yours faithfully

信函 14

致雇主，尽管收到暂停施工通知，如果未在 7 天内全额付款

敬启者：

我方在［插入日期］的信函中声明，倘若贵方未能于收函起 7 天内全额付款，我方将中止履行我方义务，目前我方仍未收到任何/全额［酌情删除］款项。特此通知贵方我方暂停履行合同规定的所有义务，此通知立即生效。

施工现场安全状况良好。但是，如果贵方坚持拒付款项，现在由贵方负责工地的安全及保险事宜。

敬上

Letter 15

To employer, requesting interest on late payment

Dear

We refer to certificate number [*insert number*] dated [*insert date*]. The final date for payment was [*insert date*]. At the time of writing we have not received any/full [*delete as appropriate*] payment.

Under the provisions of clause ____ you are obliged to pay us simple interest at 5% above Bank of England Base Rate current at the date payment became overdue until the amount is paid. However, you should note that this contractual right to interest is without prejudice to our other rights and remedies. This letter is simply by way of notice that we have no intention of waiving our right to interest although we intend to write to you under separate cover if payment is not made by close of business on the day following the date of this letter.

Yours faithfully

信函 15

致雇主，要求支付延迟付款的利息

敬启者：

我方参阅［插入日期］的［插入编号］号证书，了解到支付截止日期为［插入日期］。截止发信时，我方仍未收到任何/全额［酌情删除］款项。

根据条款____规定，贵方有义务自付款逾期时起，按照高于目前英格兰银行基准利率5%支付我方单利，直至付清款项。但是，望贵方注意，该要求获得利息的合同权利不影响我方其他权利和补救措施。该函仅在通知贵方我方无意放弃要求贵方支付利息的权利。若贵方未能于该函次日下班前支付款项，我方仍打算另函贵方。

敬上

Letter 16

To employer, requesting retention money to be placed in a separate bank account
Special delivery

Dear

We formally request you to place all current and future retention money in a separate bank account set up for the express purpose and identified as money held in trust for our benefit. Please inform us of the name of the bank, the account name and number. Notwithstanding the provisions of the contract, it is established that your obligation exists irrespective of any formal request we may make.

Yours faithfully

Copy: Engineer

信函 16

致雇主，要求将保留金存入单独银行账户
快递

敬启者：

我方现正式要求贵方将现在和今后所有的保留金存入单独银行账户，该账户设立目的明确，所存入资金是为保护我方利益的托管资金。请贵方告知我方银行名称，账户名及账号。尽管有合同条款规定，但无论我方提出何种正式要求，贵方确有义务配合。

敬上

抄送至：工程师

Letter 17

To employer, if failure to place retention in separate bank account

Special delivery

Dear

Further to our letter dated [*insert date*] you have not notified us that you have set aside retention money in a separate trust fund as we requested and as the contract and the law provides.

The separate trust fund will protect our money in the event of your insolvency. If you have not complied with our request by [*insert date*], we shall immediately seek an injunction to compel you to comply.

Yours faithfully

Copy: Engineer

信函 17

致雇主，如未将保留金存入单独账户

快递

敬启者：

参照我方［插入日期］的信函，贵方一直未通知我方，已按我方要求，按合同及法律规定，将保留金存入单独托管基金账户。

单独的托管基金将在贵方破产情况下保护我方的资金。若贵方在［插入日期］前未遵守我方要求，我方将立即寻求强制令，强制贵方遵守。

敬上

抄送至：工程师

Letter 18

To engineer，enclosing all information for preparation of final certificate

Dear

In accordance with clause ____ we enclose full details of the final account for this contract together with all supporting documentation.

We should be pleased if you would proceed with the necessary calculations and verifications to enable the final certificate to be issued in accordance with the contract timescale.

Yours faithfully

信函 18

致工程师，附上准备最终证书的所有信息

敬启者：

根据条款____，我方随函附上合同最终账目的全部详细信息及所有证明文件。

我方希望贵方继续进行必要的核算，以便能够按照合同时间表发布最终证书。

敬上

Letter 19

To engineer，enclosing final account

Dear

In accordance with clause ____ of the conditions of contract we enclose the final account and final statement referred to in clause ____ and we should be pleased to have your agreement.

We enclose the following supporting documentation：

［*List*］

If you reasonably require any further information，we will be pleased to provide it. Please let us have a complete list of such requirements (if any) within one month from the date of this letter.

Yours faithfully

信函 19

致工程师，随函附寄结算账目

敬启者：

根据合同条款____，随函附上结算账目及条款____涉及的决算表。希望贵方同意。

随函附寄以下证明文件：

［列出文件］

若贵方在合理范围内要求更多信息，我方将欣然提供。请于收函起一个月之内给我方提供完整的所需（若有的话）清单。

敬上

Letter 20

To employer, disputing the final account

Dear

We are in receipt of the employer's final account and employer's final statement.

Take this as formal notice, under the provisions of clause ____, that we dispute the entirety of such employer's final account and employer's final statement. Therefore, the balance will not become conclusive.

We suggest that a meeting is appropriate to agree the final account and we would be free to attend such a meeting on [*insert date*] at [*insert time*].

Yours faithfully

信函 20

致雇主，结算账目有争议

敬启者：

我方收到雇主最终账目和决算表。

根据条款____规定，我方正式通知贵方，我方对雇主的最终账目及决算表的总额有异议。因此，该余额将不会成为最终确定金额。

我方建议召开会议商定结算金额，我方可以在［插入日期］［插入时间］参加该会议。

敬上

Letter 21

To engineer, if final certificate not issued on the due date (a)

Special delivery

Dear

Clause ____ of the conditions of contract requires you to issue the final certificate within 2 months from the latest of the following events:

1. The end of the rectification period/last rectification period [*delete as appropriate*] – [*insert date*].

2. Date of issue of the certificate of making good/last certificate of making good [*delete as appropriate*] – [*insert date on certificate*].

3. The date on which you sent us a copy of the statement and ascertainment under clause ____ – [*insert date or, if the engineer has not sent an ascertainment and statement, add the following*:] In this instance, you have not sent us such ascertainment and statement which should have been issued on the [*insert date*].

Therefore, the final certificate should have been issued on the [*insert date calculated as above*]. Some [*insert number*] weeks have passed since that date and we have received no such certificate. You are in breach of contract, a breach for which, we are advised, the employer will become liable after receipt of this notification. Without prejudice to our rights in this matter, if the final certificate is in our hands by [*insert date*], we will take no further action on such breach.

Yours faithfully

Copy: Employer

信函 21

致工程师，如果没有如期发布最终证书（a）

快递

敬启者：

合同条款____要求贵方在最近的以下事项后的 2 个月内签发最终证书，相关事项如下：

1. 整改期/最后整改期［酌情删除］结束—［插入日期］。

2. 发出修缮证书/最后修缮证书［酌情删除］的日期—［插入开证日期］。

3. 根据条款____，贵方发给我方结算单及确认单副本的日期［插入日期，如果工程师没有寄送确认单和结算单，请添加以下内容：］。在此情况下，贵方未向我方发送本应在［插入日期］发出的结算单和确认单。

因此，最终证书本应于［插入如上计算所得的日期］发出。自该日期起已过［插入数字］周，我方仍未收到该证书。贵方已构成违约，我方获悉，在收到此通知后，雇主将对该违约承担责任。在不影响我方利益的前提下，如我方能够于［插入日期］前收到最终证书，我方将不再追究违约责任。

敬上

抄送至：雇主

Letter 22

To engineer, if final certificate not issued on the due date (b)
Special delivery

Dear

Clause ____ of the conditions of contract requires you to issue the final certificate within 28 days of the latest of the following events:

1. The sending to us of the computations of the adjusted contract sum, which we received on the [*insert date*].

2. Your certificate of making good under clause ____ - [*insert date*].

Therefore, the final certificate should have been issued on the [*insert date calculated as above*]. Some [*insert number*] weeks have passed since that date and we have received no such certificate. You are in breach of contract, a breach for which, we are advised, the employer will become liable after receipt of this notification. Without prejudice to our rights in this matter, if the final certificate is in our hands by [*insert date*], we will take no further action on such breach.

Yours faithfully

Copy: Employer

信函 22

致工程师，如果没有如期发出最终证书（b）

快递

敬启者：

合同条款____要求贵方在最近的以下事项后的28天内签发最终证书，相关事项如下：

1. 向我方发送调整后的合同金额的算法，我方在［插入日期］收到该算法。

2. 条款____要求的修缮证明—［插入日期］。

因此，最终证书本应于［插入如上计算所得的日期］发出。自该日期起已过［插入数字］周，我方仍未收到该证书。贵方已构成违约，我方获悉，在收到此通知后，雇主将对该违约承担责任。在不影响我方利益的前提下，若我方能够于［插入日期］前收到最终证书，我方将不再追究违约责任。

敬上

抄送至：雇主

Letter 23

To engineer, if final certificate not issued on the due date (c)

Special delivery

Dear

Clause ____ of the conditions of contract requires you to issue the final certificate within 28 days of receipt by you of all documentation reasonably required for computation of the amount to be certified, provided that you have issued your certificate under clause ____.

You confirmed that all such documents were in your possession by your letter of the [*insert date*]. You issued a clause ____ certificate on the [*insert date*].

Therefore, the final certificate should have been issued on the [*insert date calculated as above*]. Some [*insert number*] weeks have passed since that date and we have received no such certificate. You are in breach of contract, a breach for which, we are advised, the employer will become liable after receipt of this notification. Without prejudice to our rights in this matter, if the final certificate is in our hands by [*insert date*], we will take no further action on such breach.

Yours faithfully

Copy: Employer

信函 23

致工程师，如果没有如期发出最终证书（c）

快递

敬启者：

合同条款____要求贵方在收到计算认证金额所需的所有文件后 28 天内签发最终证书，但前提是贵方已依据合同第____条规定发出证书。

贵方在［插入日期］的函件中确认已持有以上提到的所有文件，并且于［插入日期］发出条款____所要求的证书。

因此，最终证书本应于［插入如上计算所得的日期］发出。自该日期起已过［插入数字］周，我方仍未收到该证书。贵方已构成违约，我方获悉，在收到此通知后，雇主将对该违约承担责任。在不影响我方利益的前提下，若我方能够于［插入日期］前收到最终证书，我方将不再追究违约责任。

敬上

抄送至：雇主

第八章 工程竣工与合同终止信函

第一节 工 程 竣 工 概 述

在我国，建设工程项目竣工验收是指由建设单位、施工单位和项目验收委员会，以项目批准的设计任务书和设计文件，以及国家或相关部门发布的施工验收规范和质量检验标准为依据，按照一定的程序和手续，在项目建成并试生产合格后，对工程项目的总体进行检验、认证、综合评价和鉴定的活动。《最高人民法院关于审理建设工程施工合同纠纷案件适用法律问题的解释》（法释［2004］14号）第十三条规定："建设工程未经竣工验收，发包人擅自使用后，又以使用部分质量不符合约定为由主张权利的，不予支持；但是承包人应当在建设工程的合理使用寿命内对地基基础工程和主体结构质量承担民事责任。"如果工程有质量缺陷，则属于工程保修期限内的问题，承包商通过履行保修期间的义务来完成修补。

在FIDIC合同条件下，雇主在工程基本完工，即没有全面完工的情况下，就可接受承包商的部分移交，并向承包商颁发《移交证书》。工程进入"缺陷责任期"，指工程如果还有些扫尾、清理地面等不影响工程使用的某些次要部分工程，缺陷修复工作可以在缺陷责任期内进行。全面完工后，雇主向承包商颁发《缺陷责任证书》。如果雇主要求提前接收部分已完工工程，则此部分已完工工程的相关质量责任自接收时起转至雇主。

竣工证书与雇主接收证书的发放需要相关的工程文件支撑。国际工程项目竣工文件的准备与提交是一个自然的过程，即只要有所准备，所有的与项目有关的资料，均应在项目的执行过程中按照合同规定的程序提交给工程师。FIDIC合同条件对工程实质性完工前，承包商与雇主对于工程的交接过程中，由承包商准备、整理、编制和提交的竣工文件没有专门的规定。所有的竣工文件的准备、整理、编制与提交，均在项目的实施过程中来完成，并以此作为期中付款申请和期中付款的依据之一。

本章工程竣工信函的内容主要侧重工程即将竣工或部分工程即将竣工、不当扣留竣工证书、工程早日使用等方面。

第二节 合 同 终 止 概 述

合同终止是一个严肃且具有风险的问题，合同任何一方在作出终止决定之前应审慎研究合同终止的可行性，寻求协商解决问题的可能性。本节合同终止信函侧重承包商向雇主发出的合同终止情形，以及合同终止后相关事宜的处理。

承包商终止合同需满足下列条件：

1999年FIDIC"新红皮书"、"新黄皮书"和"银皮书"规定了承包商终止合同的事项。按照"新红皮书"第16.2款的规定，如出现下列情况，承包商有权终止合同：

（1）承包商在根据承包商暂停工作的权利的规定，就雇主未能遵循雇主的资金安排规定的事项发出通知后 42 天内，仍未收到合理的证据；

（2）工程师未能在收到报表和证明文件后 56 天内签发有关的付款证书；

（3）在规定的付款时间到期后 42 天内，承包商仍未收到期中付款证书规定的应付款额；

（4）雇主实质上未能根据合同规定履行其义务；

（5）雇主未遵守第 1.6 款［合同协议书］或第 1.7 款［转让］的规定；

（6）雇主原因拖长的停工影响了整个工程；或

（7）雇主破产或无力偿债、停业清理，已有对其财产的接管令或管理令，与债权人达成和解，或为其债权人的利益在财产接管人、受托人或管理人的监督下营业，或采取了任何行动或发生任何事件（根据有关适用法律）具有与前述行动或事件相似的效果。

第三节　工程竣工信函

Letter 1

To engineer, if practical completion of the Works or a section imminent

Dear

We anticipate that the Works/section ［*delete as appropriate*］ will be complete on ［*insert date*］. If you will let us know when you wish to carry out your inspection, we will arrange for M. . . ［*insert name*］ to be on site to give immediate attention to any queries which may arise. We look forward to receiving your practical completion certificate following your inspection.

Yours faithfully

信函 1

致工程师，如果工程即将竣工或部分工程即将竣工

敬启者：

我方预计工程/部分工程［酌情删除］将于［插入日期］完工。如果贵方希望进行验收检查，请告知我方时间，我方将安排［插入名字］在现场随时解答可能出现的任何疑问。我方期待在贵方检查验收之后能够收到实际竣工证明书。

敬上

Letter 2

To engineer, if completion certificate wrongly withheld (a)

Dear

Thank you for your letter of the [*insert date*].

We are surprised to learn that, in your opinion, practical completion has not been achieved. The items you list as outstanding can only be described as trivial. We do not consider that such items can possibly justify withholding your certificate.

We strongly urge you to reconsider the matter and to issue your certificate forthwith as required by clause ____ of the conditions of contract. If we do not receive your certificate by [*insert date*] naming [*insert date*] as the date of practical completion, we will take whatever steps we deem appropriate to protect our interests.

The items you list are receiving attention in the normal way.

Yours faithfully

信函 2

致工程师，如果不当扣留竣工证书（a）

敬启者：

感谢贵方［插入日期］来函。

我方惊悉贵方认为项目没有实际完工。贵方所列的突出问题实际上微不足道。我方认为贵方没有理由因此扣留竣工证书。

我方敦促贵方重新考虑该问题，并按合同条款____的规定，立即颁发竣工证书。如未能在［插入日期］前收到竣工证书，即［插入日期］为实际竣工日期，我方将采取一切措施保护我方利益。

贵方所列问题受到正常关注。

敬上

Letter 3

To engineer, if completion statement wrongly withheld (b)

Dear

Thank you for your letter of the [*insert date*].

We are surprised to learn that, in your opinion, practical completion of the Works/section [*delete as appropriate*] has not been achieved. The items you list as outstanding can only be described as trivial and they cannot possibly justify your contention.

We draw your attention to the fact that, under this form of contract, you have no certifying function and the issue of the practical completion statement is simply a process of recording a matter of fact. It is not something for your opinion. We strongly urge you to issue the written statement forthwith as required by clause ____ of the conditions of contract. If we do not receive your statement by [*insert date*] naming [*insert name*] as the date of practical completion, we will take whatever steps we deem appropriate to protect our interests.
The items you list are receiving attention in the normal way.

Yours faithfully

信函 3

致工程师，如果不当扣留竣工证书（b）

敬启者：

感谢贵方［插入日期］来函。

我方惊悉贵方认为项目没有实际完工。贵方所列的突出问题实际上微不足道。我方认为贵方没有理由因此扣留竣工证书。

我方提请贵方注意，根据这种合同形式，贵方没有核证职能，发布实际完工说明只是记录事实的过程。贵方意见不能决定此事。按合同条款____的要求，我方敦促贵方立即发放书面竣工证明。如未能在［插入日期］前收到竣工证书，即［插入日期］为实际竣工日期，我方将采取一切措施保护我方利益。

贵方所列问题会受到正常关注。

敬上

Letter 4

To employer, consenting to early use

Dear

Further to your request dated [*insert date*] under clause ____ to use or occupy the Works before the date for completion, we have notified the insurers to seek confirmation that such use or occupation will not prejudice the insurance.

[*If insurance options B or C apply, add*:]

Subject to such confirmation, we consent to such use or occupation as you describe.

[*If insurance option A applies, add*:]

If the insurers require an extra premium, our consent will be subject to your agreement under clause ____.

Yours faithfully

信函 4

致雇主，同意早日使用

敬启者：

关于贵方［插入日期］根据条款____要求于竣工日前使用或占用工程，我方已通知承保人以确认此类使用或占用无损于保险事项。

［如果是保险选项 B 或 C，则补充：］

根据此确认，我方同意贵方所描述的使用或占用。

［如果是保险选项 A，则补充：］

如果承保人需要额外保费，我方同意与否将取决于你们双方根据条款____所达成的协议。

敬上

Letter 5

To employer, consenting to partial possession (a)

Dear

In response to your letter of the [*insert date*], we consent to your request to take partial possession of the Works, namely [*describe part or parts*] provided:

1. The date for possession will be [*insert date*] and the engineer will give a written statement to that effect on your behalf in accordance with clause ____.
2. [*Insert whatever particular conditions may be appropriate to the circumstances.*]

If you will, or you will authorise the engineer to, write to us indicating agreement to the above conditions, we will make arrangements to hand over the appropriate keys on the [*insert date*].

Yours faithfully

Copy: Engineer

信函 5

致雇主，同意部分接收（a）

敬启者：

回复贵方［插入日期］来函，我方同意贵方接收部分工程的请求，即可接收［描述部分或相关部分］：

1. 接收日期为［插入日期］，工程师将根据条款____代表贵方出具书面说明。

2. ［插入一切可适用于此情况的详细条件。］

如果贵方或贵方授权工程师，书面表明同意以上条款，我方将安排于［插入日期］移交相关钥匙。

敬上

抄送至：工程师

Letter 6

To employer，consenting to partial possession (b)

Dear

In response to your letter of the [*insert date*]，we consent to your request to take partial possession of the Works，namely [*describe part or parts*] provided：

1. The date for possession will be [*insert date*].
2. [*Insert whatever particular conditions may be appropriate to the circumstances.*]

If you will write to us agreeing to the above conditions，we will make the necessary arrangements to hand over the appropriate keys on the [*insert date*]. We shall then issue a written statement in accordance with clause ____.

Yours faithfully

信函 6

致雇主，同意部分接收（b）

敬启者：

回复贵方［插入日期］来函，我方同意贵方接收部分工程的请求，即可接收［描述部分或相关部分］：

1. 接收日期为［插入日期］。

2. ［插入一切可适用于此情况的详细条件。］

如果贵方书面同意以上条款，我方将做必要的安排，于［插入日期］移交相关钥匙。然后，我方将根据条款____发布书面声明。

敬上

Letter 7

To employer，issuing written statement of partial possession

Dear

Further to your letter of the [*insert date*] indicating agreement to the conditions contained in our letter of the [*insert date*]，take this as the written statement which we are to issue in accordance with clause ____ of the conditions of contract. We identify the part (s) of the Works taken into possession (the relevant part (s)) as [*describe the part or parts in sufficient detail to allow no mistake*]. The date on which you took possession (the relevant date) was [*insert date*].

We draw your attention to the consequences of partial possession，particularly as they apply to the commencement of the rectification period，your insurance liability and the reduction in any liability which we may have for liquidated damages.

Yours faithfully

信函 7

致雇主，发布部分接收书面声明

敬启者：

贵方［插入日期］来函表明同意我方［插入日期］函中所提条件，我方根据合同条款____发布书面声明。我方确定项目工程被接收部分（相关部分）为［描述部分或相关部分的具体细节，不要出错］。贵方接收日期（相关日期）为［插入日期］。

我方提请贵方注意部分接收的影响，尤其是影响整改期的开始，贵方的保险责任，以及减小违约赔偿责任中的我方责任。

敬上

Letter 8

To employer，refusing consent to partial possession

Dear

Thank you for your letter of the [*insert date*] requesting our consent to you taking partial possession of the Works，namely [*describe part or parts*].

We regret that we feel unable to give our consent in this instance，because [*insert reasons*].

[*If appropriate*，*add*:]

We will let you know immediately if circumstances change so substantially that we feel able to consent to your request.

Yours faithfully

Copy：Engineer [*omit when using DB*]

信函 8

致雇主，拒绝同意部分接收

敬启者：

收悉［插入日期］来函，函中要求我方同意贵方接收部分工程，即［描述部分或相关部分］。

在此情况下我方不能同意，为此深表遗憾，因为［插入原因］。

［若合适，则补充：］

如果情况发生显著变化，致使我方认为能够同意贵方请求，我方会立即告知贵方。

敬上

抄送至：工程师［使用 DB 模式时删除］

Letter 9

To engineer, after receipt of schedule of defects

Dear

Thank you for your instruction number [*insert number*] dated [*insert date*] scheduling the defects you require to be made good now that the rectification period has ended.

We have carried out a preliminary inspection and we are making arrangements to make good most of the items on your schedule. However, we do not consider that the following items are our responsibility for the reasons stated:
[*List, giving reasons.*]

We shall, of course, be happy to attend to such items if you will let us have your written agreement to pay us daywork rates for the work.

Yours faithfully

信函 9

致工程师，收到建筑瑕疵细目表后

敬启者：

感谢贵方［插入日期］发出指令号［插入编号］，要求修缮贵方列出的建筑瑕疵，目前整改阶段已结束。

我方已进行预先检查，并正在安排修缮细目表所列大多数事项。但我方不认为下述项目由我方负责，理由如下：

［列表，提供理由。］

当然，如果贵方书面同意按日支付工程款项，我方将欣然处理此类事项。

敬上

Letter 10

To engineer, when making good of defects completed

Dear

We are pleased to inform you that all making good of defects has been completed in accordance with your schedule [*if appropriate, substitute 'your amended schedule'*]. We should be pleased if you would carry out your own inspection and confirm your satisfaction by issuing a certificate of making good.

Yours faithfully

信函 10

致工程师，修缮工作完成时

敬启者：

我方很高兴通知贵方我方已按贵方进度［如适当，用“贵方修订后的进度表”替代］完成所有修缮工作。如果贵方亲自检查，并颁发修缮完好证书以确认贵方满意，我方不胜感激。

敬上

Letter 11

To engineer, returning drawings, etc. after final payment if requested

Dear

Thank you for your letter of [*insert date*].

We enclose, as requested, all copies of drawings, details, descriptive schedules and other documents of like nature which bear your name and which are in our possession. We have, naturally, retained our copy of the contract documents for record purposes.

Yours faithfully

信函 11

致工程师，如果要求，可于最终付款后退还图纸等

敬启者：

收悉贵方［插入日期］来函。

按要求随函附寄所有图纸副本，细目表副本，描述性进度表副本和其他类似标有贵方名称且在我方手中的文件副本。当然，我方已保存合同文件副本用作备案。

敬上

第四节　合同终止信函

Letter 1

To employer or engineer, if default notice served

Special delivery

Dear

We are in receipt of your letter of the [*insert date*], which apparently you intend to be a default notice in accordance with clause ____ of the conditions of contract.

[*Add either*:]

The purported notice contains serious error.

[*Or*:]

We are advised that your purported notice is ambiguous.
[*Or*:]
The substance of the default specified in your purported notice is incorrect.
[*Or*:]
The default specified in your purported notice is the result of your own default. [*Explain as appropriate*.]
[*Or*:]
Your purported notice is wrongly served.

[*continued*]

信函 1
致雇主或工程师，如果有违约通知
快递

敬启者：

收悉［插入日期］来函，显然贵方打算根据合同条款____将此作为违约通知发给我方。
［补充：］

通知中存在严重错误。
［或者：］
我方认为贵方声称的通知模棱两可。

［或者：］
贵方通知中所指的违约内容不正确。
［或者：］
通知中所指的违约缘于贵方自己违约。［请合理解释］
［或者：］
贵方提供通知有误。

待续

Letter 1 continued

[*Th4en add*:]

Your notice is, therefore, invalid and of no effect. Take this as notice that if you proceed to give notice of termination, it will be unlawful, it will amount to repudiation and we will take immediate proceedings against you/the employer [*delete as appropriate*].

Yours faithfully

Copy: Engineer/Employer [*delete as appropriate*]

信函1 续:

[然后补充:]

因此，贵方通知无效。望贵方注意，若继续发出终止通知，将属违法且等同于毁约。对此，我方将立即对贵方/雇主［酌情删除］提起诉讼。

敬上

抄送至：工程师/雇主［酌情删除］

Letter 2

To employer or engineer, if default notice served justly

Special delivery

Dear

We are in receipt of your letter of the [*insert date*] which you sent as a default notice in accordance with clause ____ of the conditions of contract.

We regret that you have felt it necessary to send such a notice, but we are pleased to be able to inform you that [*insert whatever steps are being taken to remove the default*].

Yours faithfully

Copy: Engineer/Employer [*delete as appropriate*]

信函 2

致雇主或工程师，如果恰当提供违约通知

快递

敬启者：

收悉［插入日期］来函，贵方根据合同条款____发出此违约通知。

贵方认为有必要发布该通知，我方对此深表遗憾，但是很高兴告知贵方［插入任何为消除违约行为而采取的措施］。

敬上

抄送至：工程师/雇主［酌情删除］

Letter 3

To employer, if premature termination notice issued

Special delivery

Dear

We are in receipt of your notice dated [*insert date*] purporting to terminate our employment under clause ____ of the conditions of contract.

Your original default notice was received on the [*insert date*]. The Post Office will be able to confirm to you the date of delivery. Your notice of termination was, therefore, premature and of no effect and may amount to a repudiation of the contract for which we can claim substantial damages. We have already corrected the default specified in your original notice and, without prejudice to any of our rights and remedies in this matter and particularly (but without limitation) our right to treat your purported termination as repudiation, we will continue to work normally while we take appropriate advice.

Yours faithfully

信函 3

致雇主，如果提前发布终止通知书

快递

敬启者：

收悉贵方［插入日期］通知书，贵方打算依照合同条款____终止我方的雇用关系。

我方［插入日期］收到贵方最初的违约通知书，邮局将能向贵方确认递送日期。因此，贵方的终止通知书发出过早且无效，而且此通知可能等同于废除合同，我方可以对此要求巨额赔偿。我方已经更正了贵方最初通知中指出的违约行为，但这不影响我方在本事项中的任何权利和补救措施，特别是（但不限于）我方将贵方声称的终止视为废除合同的权利，我方将继续正常施工，同时咨询律师意见。

敬上

Letter 4

To employer, who terminates after notification of cessation of terrorism cover

Dear

We are in receipt of your termination notice given under clause ____. You should note that, although clause ____ appears to preclude further payment, it is now established that it does not remove your obligation to pay, by the final date for payment, any amount due.

Yours faithfully

信函 4

致雇主，雇主在公布恐怖主义停止后发出终止通知

敬启者：

收到贵方根据条款____发出的终止通知。望贵方注意，尽管条款____似乎避免进一步付款，但现在事实证明该条款并不排除贵方在最后付款日期前支付任何应付金额的义务。

敬上

Letter 5

To employer, giving notice of default before termination
Special delivery

Dear

We hereby give you notice under the provisions of clause ____ of the conditions of contract that you are in default/a specified suspension event has occurred [*delete as appropriate*] in the following respect:
[*Insert details of the default with dates if appropriate and refer to the appropriate sub-clause.*]
which must be ended.

If you continue the default/specified suspension event [*delete as appropriate*] for 14 days after receipt of this notice, we may forthwith terminate our employment under this contract without further notice.

Yours faithfully

信函 5

致雇主，于终止雇用关系前发出违约通知书
　　快递

敬启者：

　　根据合同条款____规定，特此通知贵方，在如下方面，贵公司存在违约/出现具体停工事件［酌情删除］：

　　［如合适，请补充附带日期的违约详情，并参阅相应的子条款］提及事项必须终止。

　　如在收到通知后 14 天内，贵方继续违约/继续出现具体停工事件［酌情删除］，我方可能立即终止本合同下的雇用关系，恕不另行通知。

　　敬上

Letter 6

To employer, terminating employment after default notice

Special delivery

Dear

We refer to the default notice sent to you on the [*insert date*].

Take this as notice that, in accordance with clause ____, we hereby terminate our employment under this contract without prejudice to any other rights or remedies which we may possess.

We are making arrangements to remove all our temporary buildings, plant etc., and materials from the site and we will write to you again within the next week regarding financial matters.

Yours faithfully

信函6

致雇主，在收到违约通知后解除雇用关系

快递

敬启者：

我方参考［插入日期］发送给贵方的违约通知。

根据第____条款，通知贵方，我方特此终止本合同下我们双方的雇用关系，同时不影响我方可能拥有的任何其他权利或补救措施。

目前我方正在安排从现场移除所有临时建筑、设备、材料等。相关财务事宜将于下周函告贵方。

敬上

Letter 7

To employer, terminating employment on the employer's insolvency

Special delivery

Dear

In accordance with the provisions of clause ______ of the conditions of contract, take this as notice that we hereby terminate our employment, because [*insert precise details of the insolvency event*]. This termination will take effect on receipt of this notice.

We are making arrangements to remove all our temporary buildings, plant etc., and materials from the site and we will write to you again within the next week regarding financial matters. This notice is without prejudice to any other rights and remedies we may possess.

Yours faithfully

信函 7

致雇主，雇主破产即解除雇用关系

快递

敬启者：

根据合同条款____，通知贵方，我方特此终止我们双方的雇用关系，原因是［补充资不抵债事项的确切细节］。收到此通知时，终止协议即生效。

目前我方正在安排从现场移除所有的临时建筑、设备、材料等。相关财务事宜将于下周函告贵方。该通知不影响我方可能拥有的任何其他权利或补救措施。

敬上

Letter 8
To employer，where either party may terminate
Special delivery

Dear

The carrying out of the whole or substantially the whole of the uncompleted Works has been suspended for a continuous period of [*insert the relevant period from the contract particulars*] by reason of the following event（s）：[*insert details of the event*（*s*）].

Take this as notice，in accordance with the provisions of clause ___ of the conditions of contract，that unless the suspension ceases within 7 days of receipt of this notice，we may terminate our employment under the contract.

Yours faithfully

信函 8
致雇主，任何一方均可终止合约的情形
快递

敬启者：

整个或大部分未完工工程的施工已经持续中止一段时期［插入合同细节规定的相关时期］，原因如下：［补充事项详情］。

根据合同条款___，我方通知贵方如果在收到该通知起七天内不结束暂停施工，我方可依照合同解除我们双方的雇用关系。

敬上

Letter 9

To employer, terminating if suspension has not ceased after notice

Special delivery

Dear

Further to our notice served in accordance with the provisions of clause ____, the suspension has not ceased and we now hereby terminate our employment under the contract.

The consequences of such termination are set out in clause ____.

Yours faithfully

信函 9

致雇主，如果收到通知后暂时停工仍未停止，终止雇用关系

快递

敬启者：

正如我方依据条款____发给贵方的通知内容所述，由于暂时停工仍未停止，我方现特此终止本合同下的双方雇用关系。

在条款____中，列有此类终止的后果。

敬上

Letter 10

To employer，terminating employment after damage by insured risk

Special delivery

Dear

We refer to our notice sent to you under the provisions of Schedule ____，paragraph ____ on [*insert date*]. We consider that it is just and equitable to terminate our employment in accordance with paragraph ____ and we hereby exercise our option to so terminate forthwith. [*The termination must be received by the employer not less than 28 days from the date of the occurrence of the loss or damage.*]

Yours faithfully

信函 10

致雇主，出险造成损失后终止雇用关系

快递

敬启者：

我方参考［插入日期］发给贵方的通知，该通知依据段落____，附表____的规定发出。我方认为根据段落____，终止双方雇用关系是公平公正之举，我方谨此行使我方的选择权，立即终止。［自该损失或损害发生之日起计不少于28天内，雇主必须收到终止通知。］

敬上

第九章　分包合同信函

第一节　分包合同概述

一、分包含义

分包（ subcontracting）是指主承包商将部分工程交由他人实施和完成的行为。在分包合同关系中，分包商只是承揽、实施和完成主承包商交给他的部分工程，而主合同中对雇主的全部责任和义务仍由主承包商承担。分包合同成立的前提是雇主与主承包商签订主合同。没有雇主和主承包商签订的主合同为前提，分包就不能成立。

英国 ICE 第七版第 59（3）条有关主承包商对指定分包的责任明确规定：“除非本条款和第 58（3）条另有规定，主承包商应对他雇用的指定分包商所做的工程、提供的货物材料或服务负责，如同主承包商自己实施这些工程，提供这些货物材料、服务一样。”英国 NEC 合同条件没有指定分包商的条款，他们认为工程施工合同的原则是总承包商对其承包工程的各方面管理负全责，对分包商的指定会与此原则相矛盾，而且会引起许多实际问题。

在国外，发达的分包体系已成为国外建筑业的特点之一。国外的大型工程承包公司同国内公司相比，管理人员比例较高。某些主承包商企业甚至是纯粹的管理型企业，管理人员素质很高，他们在承担项目时，会将所有的具体施工任务分包出去，自己则专门从事项目管理工作，专业划分详细而全面。

本章分包合同信函主要侧重请求雇主同意转让、分包、意向书、担保、分包商、提名分包商、指定分包商管理等方面内容的介绍。

二、主承包商与分包商之间的关系特点

（一）主承包商与分包商之间的责任传递

国际工程合同主承包商与分包商关系的一个最大特点就是两者之间就分包工程产生权利义务的责任传递。承包商应将分包商、分包商的代理人或雇员的行为或违约视为承包商自己的行为或违约，并为之承担全部责任。

（二）工程师对分包商的认可

根据 FIDIC 合同规定，承包商在选择材料供应商或向合同中已注明的分包商进行分包时，无需征得工程师同意；但是其他拟雇用的分包商须得到工程师的事先同意；而承包商应至少提前 28 天将每位分包商的工程预期开工日期以及现场开工日期通知工程师。

（三）承包商对指定分包商的反对

根据 FIDIC“新红皮书”规定，“指定分包商”是指一个分包商：（1）合同中指明作为指定分包商的，或（2）工程师依据第 13 条［变更和调整］指示承包商将其作为一名分包商雇用的人员。

承包商没有义务雇用一名他已通知工程师并提交具体证明资料说明其有理由反对的指定分包商。如果因为下述任何事宜（包括但不限于）而反对，则该反对应被认为是合理的，除非雇主同意保障承包商免于承担下述事宜的后果：

(1) 有理由相信分包商没有足够的能力、资源或资金实力；

(2) 分包合同未规定指定分包商应保障承包商免于承担由分包商、其代理人、雇员的任何疏忽或对货物的错误操作的责任；或

(3) 分包合同未规定指定分包商对所分包工程（包括设计，如有时），应该：

1) 向承包商承担该项义务和责任以使承包商可以依照合同免除他的义务和责任，以及；

2) 保障承包商免于按照合同或与合同有关的以及由于分包商未能履行这些义务或完成这些责任而导致的后果所具有的所有义务和责任。

(四) 主承包商对指定分包商的支付

承包商应向指定分包商支付经工程师证实的根据分包合同应支付的款额。在颁发一份包括支付给指定分包商的款额的支付证书之前，工程师可以要求承包商提供合理的证据，证明按以前的支付证书已向指定分包商支付了所有应支付的款额。除非承包商向工程师提交了合理的证据，或以书面材料使工程师同意他有权扣留或拒绝支付该项款额，以及向工程师提交了合理的证据表明他已将此权力通知了指定分包商，否则，雇主应（自行决定）直接向指定分包商支付部分或全部已被证实应支付给他的款额。

第二节 分包合同信函

Letter 1

To employer, requesting consent to assignment

Dear

In accordance with clause ____ we should be pleased to receive your consent to the assignment of our rights to payment under this contract to [*insert name*].

We wish to take such action, because [*give reasons briefly*]. We acknowledge that our obligations under the contract will be unaffected.

Yours faithfully

信函 1

致雇主，要求同意转让

敬启者：

根据合同条款____，我方希望贵方同意我方将本合同项下我方的付款权转让给［插入名称］。

我方希望采取此行为，因为［插入原因］。我方确认我方在本合同下的义务不会受到影响。

敬上

Letter 2

To sub-contractor，assessing competence under the CDM Regulations[1]

Dear

In accordance with Regulation 4 of the CDM Regulations 2007，we are obliged to take steps to satisfy ourselves that you are competent. [*Where the contractor and sub-contractor have a long and on-going business relationship add*：] Although，in your case，this is little more than a formality we have to show that we have taken reasonable steps. [*Then continue*：]

The Approved Code of Practice：'Managing health and safety in construction' states that competence should be assessed in relation to certain core criteria which have been agreed by industry and HSE. We attach a copy of these criteria which form Appendix 4 of the Code. We should be grateful if，by the [*insert date*]，you would furnish the evidence which we have highlighted in the third column. The Code states that 'unnecessary bureaucracy associated with competency assessment obscures the real issues . . . ' and we have endeavoured to take this into account when requesting evidence.

Alternatively，if you are subject to an independent accreditation organisation which assesses competence against these criteria，please provide relevant details.

Yours faithfully

---

[1] CDM Regulations：The Construction (Design and Management) Regulations 2015 (CDM 2015) came into force on 6 April 2015，replacing CDM 2007. This publication provides guidance on the legal requirements for CDM 2015 and is available to help anyone with duties under the Regulations. It describes：the law that applies to the whole construction process on all construction projects，from concept to completion；what each duty holder must or should do to comply with the law to ensure projects are carried out in a way that secures health and safety. CDM 2015 is subject to certain transitional provisions which apply to construction projects that start before the Regulations come into force and continue beyond that date.

信函 2

致分包商，根据《建筑设计管理规定》进行能力评估

敬启者：

按照《2007 年建筑设计管理规定》中的第四项内容，我方必须按照规定来评估贵方具有满足我方要求的能力。[承包商和分包商有长期且持续的业务关系时补充：] 虽然对于贵方来讲，这只是一种形式，但我方必须证明我方按照合理的步骤行事。

[然后继续：]

《经批准的实施准则》中的“健康安全施工管理条例”明确指出应按照工业和健康安全管理局所认可的核心标准来评估能力。我方附上一份标准，这些标准在《实施准则》附录 4 中。我方希望贵方在 [插入日期] 前提供合同第三卷中强调的证明。《实施准则》指出“与能力评估相关的不必要官僚作风掩盖真正的问题……”。我方在要求贵方出具证明时会尽力考虑这一点。

此外，如果贵方需要独立的认证机构依据这些标准评估贵方的能力，请提供相关详细信息。

敬上

Letter 3
To engineer, requesting consent to sub-letting

Dear

We propose to sub-let portions of the Works as indicated below, because [*state reasons*]. We should be pleased to receive your consent in accordance with clause ____.

[*List the portions of the Works or design and the names of the sub-contractors.*]

Yours faithfully

信函 3
致工程师，要求同意转包

敬启者：

我方建议转包工程的以下部分，因为［插入原因］。根据条款____，希望贵方同意。

［列举工程或设计的内容以及分包商的名称。］

敬上

Letter 4

To sub-contractor：letter of intent

Special delivery

Dear

Your quotation of the [*insert date*] for [*insert nature of the work or design*] is acceptable and we intend to enter into a sub-contract with you after the main contract documents have been satisfactorily executed.

It is not our intention that this letter，taken alone or in conjunction with your quotation，should form a binding contract. However，we are prepared to instruct you to [*insert the limited nature of the work or design required*]. If，for any reason，the project does not proceed or we instruct you to cease work，our commitment will be strictly limited to payment for the properly executed work you have completed at our request up to the date of our notification that the project will not proceed and/or our instruction to you to cease work. The basis of payment will be the prices in your quotation noted above.

No other work [*substitute 'design' if appropriate*] included in your quotation must be carried out without a further written order. No further obligation is placed upon us under any circumstances.

Yours faithfully

信函 4

致分包商：意向书

快递

敬启者：

我方接受贵方［插入日期］对［插入工程或设计内容］的报价。我方打算在总包合同文件圆满签署后与贵方签订分包合同。

我方不打算将该意向书，单独或与贵方的报价一起，形成具有约束力的合同。但是，我方准备要求贵方承担［插入局部工程或所需设计］。如果由于任何原因，项目无法继续或我方指示贵方停止施工，我方所承担的义务仅限于支付贵方在我方通知贵方项目无法继续，或停止施工的日期为止，贵方按照我方要求正确完成的施工量。付款依据将是上述提到的贵方报价中的价格。

如果没有进一步的书面指示，不得进行贵方报价中包含的任何其他施工工作［替代为“设计”，如果适当］。在任何情况下，我方不承担任何其他义务。

敬上

Letter 5

To sub-contractor，regarding part of the construction phase plan

Dear

We draw your attention to those parts of the construction phase plan which are applicable to the sub-contract works and annexed to the schedule of information. The purpose of the plan is to show the way in which the construction phase is to be managed and the important health and safety issues in this project.

Please note that the plan is not to be treated as a mere paper exercise. Rather，it is an important tool in the construction process and something which is a mandatory requirement under the Construction（Design and Management）Regulations 2007.

Copies of the complete plan have been sent to the client and other consultants and relevant parts of the plan to other parties as necessary.

Yours faithfully

Copy：CDM Co-ordinator [*unless the contractor takes this role*]

信函 5

致分包商，关于部分施工阶段计划

敬启者：

我方提请贵方注意施工阶段计划中适用于分包合同工程并附加在信息进度表中的那些部分。该计划旨在说明施工阶段管理的方式以及本项目中重要的健康和安全问题。

请贵方注意切不可视该计划为一纸空文。相反，该计划是施工过程中的重要工具，并且是《2007 年建筑设计管理规定》的强制性要求。

完整计划的副本已经发送给当事人和其他顾问，并且计划的相关部分也已发给其他必要人员。

敬上

抄送至：CDM 协调员［除非承包商承担此角色］

Letter 6

To sub-contractor，enclosing part of the construction phase plan

Dear

We enclose，for your attention，those parts of the construction phase plan which are applicable to the sub-contract works. The purpose of the plan is to show the way in which the construction phase is to be managed and the important health and safety issues in this project.

Please note that the plan is not to be treated as a mere paper exercise. Rather，it is an important tool in the construction process and something which is a mandatory requirement under the Construction（Design and Management）Regulations 2007.

Copies of the complete plan have been sent to the client and other consultants and relevant parts of the plan to other parties as necessary.

Yours faithfully

Copy：CDM Co-ordinator [*unless the contractor takes this role*]

信函 6

致分包商，附上部分施工阶段计划

敬启者：

请贵方注意，我方随函附上施工阶段计划中适用于分包合同工程的部分。该计划旨在说明施工阶段管理的方式以及本项目中重要的健康和安全问题。

请贵方注意切不可视该计划为一纸空文。相反，该计划是建筑施工过程中的重要工具，并且是《2007 年建筑设计管理规定》的强制性要求。

完整计划的副本已经发送给当事人，其他顾问，并且计划的相关部分也已发给其他必要人员。

敬上

抄送至：CDM 协调员 [除非承包商承担此角色]

Letter 7

To domestic sub-contractor, requiring a warranty if not noted in the invitation to tender

Dear

We refer to your quotation dated [*insert date*] in the sum of [*insert amount*] for [*insert the nature of the work or design*].

We are prepared to enter into a sub-contract with you if you will confirm in writing that you are prepared to sign the attached warranty/execute the attached warranty as a deed [*delete as appropriate*] in favour of the employer/future tenants/purchasers/funders [*delete as appropriate*] within two days of receipt of our instruction to do so. This is a requirement of the employer under the main contract and it cannot be varied. If you are not prepared to give the undertaking we seek, we shall have no alternative but to place the sub-contract work elsewhere.

We look forward to receiving your confirmation by [*insert date*] so that we may proceed with the contract documentation.

Yours faithfully

信函 7

致国内分包商，如果在招标邀请函中未注明，则在此要求提供担保书

敬启者：

我方参考贵方［插入日期］对［插入工程或设计性质］工程的报价［插入金额］。

如果贵方在收到我方指示 2 日内，以书面形式确认准备签署所附担保书/立契约执行所附担保书，同意雇主/未来承租人/购买者/投资人的要求，我方准备与贵方签订分包合同。这是主承包合同下雇主的要求，不得更改。如果贵方不准备提供我方寻求的担保，我方别无选择，只能将分包工程转包他处。

我方期待在［插入日期］前收到贵方确认书，以便我方可以着手处理合同文件。

敬上

Letter 8

To domestic sub-contractor, requiring a warranty if not noted in the contract documents

Dear

We have received notice from the employer that a warranty is required in favour of [*insert name*], the employer/tenant/purchaser/funder [*delete as appropriate*] and we enclose the relevant warranty.

Please sign/execute as a deed [*delete as appropriate*] and return to us with the contract documents. This is exactly the same form of warranty which was attached to the sub-contract documents.

Please let us have the completed warranty by [*insert date*]. As soon as we receive a copy of the completed warranty from the employer we will send you a copy for your records.

Yours faithfully

信函 8

致国内分包商，如果合同文件中未注明，则在此要求提供担保书

敬启者：

我方收到雇主通知要求贵方提供保证书，同意［插入名称］，雇主/承租人/购买者/投资人［酌情删除］的要求。我方附上相关保证书。

请签署/立契约执行保证书［酌情删除］，并随合同文件一并返还给我方。该保证书的格式与附在分包合同中的保证书格式完全相同。

我方希望在［插入日期］前收到签署完毕的保证书。我方收到雇主签署的保证书副本后会立即向贵方发送一份副本以供存档。

敬上

Letter 9

To engineer, if domestic sub-contractor refuses to provide a warranty which was not originally requested

Dear

[*Insert name of sub-contractor*] refuses to enter into a warranty on the terms which you have proposed.

[*Add either*:]

It appears that this sub-contractor is not prepared to enter into any warranty. Our problem is, as you are aware, that the number of sub-contractors that are able to carry out this kind of work/design [*delete as appropriate*] is extremely limited and this contractor is in great demand. We should be pleased to have your observations and instructions.

[*Or*:]

It appears that this sub-contractor is prepared to enter into a warranty if the terms are amended and we enclose an example of such a warranty. We should be glad to have your confirmation that the amended terms are satisfactory.

Yours faithfully

信函 9

致工程师，如果国内分包商拒绝签订最初没有要求的保证书

敬启者：

[插入分包商的名称] 拒绝按照贵方建议条款签订保证书。

[要么补充：]

此分包商似乎不准备签订任何保证书。如贵方所知，我方的问题是能够进行这种施工/设计 [酌情删除] 的分包商数量极其有限，并且此承包商在市场上非常抢手。我方期待收到贵方的意见和指示。

[或者补充：]

如果修改保证条款，此分包商准备签署保证书。我方随函附上一份保证书范本。我方希望贵方确认修改过的条款令人满意。

敬上

Letter 10

To engineer, objecting to a named person

Dear

We are in receipt of your instruction number [*insert number*] dated [*insert date*] instructing us to enter into a sub-contract with [*insert name*].

We have reasonable objection, under ____, to entering into such sub-contract. The reason for our objection is [*explain*].

Yours faithfully

信件 10

致工程师，拒绝提名分包商

敬启者：

我方收到贵方［插入日期］［插入编号］的指示，要求我方与［插入姓名］签署分包合同。

根据合同条款____规定，我方有理由拒绝签署此分包合同。我方拒绝的理由是［插入理由］。

敬上

Letter 11

To engineer, objecting to a nominated sub-contractor

Dear

We are in receipt of your instruction number [*insert number*] dated [*insert date*] instructing us to enter into a sub-contract with [*insert name*] for [*insert nature of work*] work.

We have reasonable objection to the employment of such nominated sub-contractor, because [*state reasons*]. This objection is made under the provisions of clause ____ of the conditions of contract.

Yours faithfully

信函 11

致工程师，拒绝指定分包商

敬启者：

我方收到贵方［日期］［编号］的指示，要求我方就［插入工程性质］工程与［插入姓名］签署分包合同。

我方有理由拒绝雇用该指定分包商，原因是［说明原因］。拒绝此事的依据是合同条款____。

敬上

Letter 12

To engineer, if contractor unable to enter into a sub-contract with named person

Dear

In accordance with supplemental provision ____, we have attempted to enter into a sub-contract with [*insert name*] that was named in the Employer's Requirements at [*insert reference to page and item number*]. Our efforts have been unsuccessful, because [*insert reason*] and we should be pleased if you would operate the provisions of paragraph ____ and either:

1. Issue a change instruction to amend the item in the Employer's Requirements so that we can enter into the sub-contract; or

2. Issue a change instruction to omit the named sub-contract work and issue further instructions about the carrying out of that work.

Yours faithfully

信函 12

致工程师，如果承包商无法与提名人签订分包合同

敬启者：

根据补充条款____，我方曾试图与"雇主要求"中［插入引用页页码和条款编号］提名人［插入姓名］订立分包合同。我方的努力并未成功，因为［插入原因］。我方希望贵方能够执行条款____中的内容，两者之一：

1. 发布修改指令，修改"雇主要求"中的条款，以便我方能签订分包合同；或者

2. 发布修改指令，取消提名分包工程，并且发布关于实施该工程的其他指示。

敬上

Letter 13

To engineer, if unable to enter into sub-contract with named person in accordance with particulars

Dear

We hereby notify you in accordance with ____ of the conditions of contract that we are unable to enter into a sub-contract with [*insert name*] in accordance with the particulars given in the contract documents. The following are the particulars which have prevented the execution of such sub-contract:

[*Specify particulars.*]

We should be pleased if you would issue your instructions as required under the contract.

Yours faithfully

信件 13

致工程师，如无法按照合同细则与提名人订立分包合同

敬启者：

根据合同条款____要求，我方特此通知贵方，依据合同文件中的细则，我方无法与[插入名字]签署分包合同。以下是阻碍执行此分包合同的细则：

[明确细则。]

我方希望贵方按照合同要求发布指示。

敬上

Letter 14

To engineer, if contractor enters into a sub-contract with named person

Dear

This letter is to notify you, as required by supplemental provision ____, that we entered into a sub-contract with [*insert name*] on [*insert date*].

Yours faithfully

信函 14

致工程师，如果承包商与提名人签订分包合同

敬启者：

根据补充条款____的要求，本函旨在通知贵方我方在［插入日期］与［插入名称］签订了分包合同。

敬上

Letter 15

To sub-contractor，regarding insurance

Dear

Please submit insurance policies and premium receipts in respect of the insurance which you are required to maintain under clause ____ of the sub-contract. The policies and receipts must be in our hands by [*insert date*].

Yours faithfully

信函 15

致分包商，关于保险

敬启者：

请提交分包合同条款____要求贵方续保的保单及保费收据。我方必须在［插入日期］前收到保单和保费收据。

敬上

Letter 16

To sub-contractor that fails to maintain insurance cover

Dear

We refer to our telephone conversation today with your M... [*insert name*] and confirm that you are unable to produce documentary evidence that the insurance required by clause ____ of the sub-contract has been properly effected and maintained.

In view of the importance of the insurance and without prejudice to your liabilities under clauses ____, we are arranging to exercise our rights under clause ____ immediately. Any sum or sums payable by us in respect of premiums will be deducted from any money due or to become due to you or will be recovered from you as a debt.

Yours faithfully

信函 16

致未参保的分包商

敬启者：

今日我方与贵方经理［插入姓名］通话，确认贵方不能出示分包合同条款____中要求的保险已正式生效并续保的文件证明。

鉴于保险的重要性，并且不损害贵方根据条款____所承担的责任，我方正在安排立即行使条款____规定的我方权利。我方就保费支付的任何款项将从任何应付或应付给贵方的款项中扣除，或将作为债务从贵方收回。

敬上

Letter 17

To sub-contractor, enclosing drawings

Dear

In accordance with clause ____ of the sub-contract, we enclose two copies of each of drawings numbered [*insert numbers*] which we consider to be reasonably necessary to enable you to carry out and complete the sub-contract works.

Yours faithfully

信函 17

致分包商，随函附上图纸

敬启者：

根据分包合同条款____规定，我方随函附上我方认为必要的编号为［插入编号］的图纸各两份以便贵方能够实施并完成分包工程。

敬上

Letter 18

To sub-contractor that sub-lets without consent

Dear

We are informed that you have sub-let [*insert part of the works sub-let*] to [*insert name*].

Since we have not given our consent, your action is in breach of the sub-contract clause ____ and must cease forthwith. Please confirm, by return, that you will comply with this letter otherwise we will take action to terminate your employment under clause ____.

Yours faithfully

信函 18

致未经同意而转包工程的分包商

敬启者：

我方获悉贵方已将[插入分包工程部分]转包给[插入名称]。

因未经我方同意，贵方的行为已违反了分包合同条款____，贵方必须立刻停止违约。请确认并回复贵方将遵守本函要求，否则我方将根据条款____终止与贵方的雇用关系。

敬上

Letter 19
To sub-contractor, giving consent to sub-letting

Dear

In response to your letter of the [*insert date*], we are pleased to give our consent, under the provisions of clause ____ of the sub-contract, to the sub-letting of [*insert portion of the works*] to [*insert name*].

Yours faithfully

信函 19
致分包商，同意转包

敬启者：

回复贵方［插入日期］来函，根据分包合同条款____，我方同意贵方将［插入转包工程部分］转包给［插入名称］。

敬上

Letter 20

To sub-contractor, requiring compliance with direction

Special delivery

Dear

Take this as notice under clause ____ of the conditions of sub-contract that we require you to comply with our direction number [*insert number*] dated [*insert date*], a further copy of which is enclosed.

If within 7 days of receipt of this notice you have not begun to comply, we will employ and pay others to comply with such direction. An appropriate deduction, which in this instance will amount to all costs incurred in connection with such employment, will be taken into account in the calculation of the final sub-contract sum or will be recovered from you as a debt.

Yours faithfully

Copy: Engineer

信函 20

致分包商，要求遵守指示

快递

敬启者：

根据分包合同条款____规定，我方特此通知贵方并要求贵方遵守我方于［插入日期］发出的编号［插入编号］的指示，并附上其另一副本。

如果在收到本通知 7 日内，贵方还未开始遵照指示开展施工，我方将雇用其他遵照该指示的人员施工并支付其费用。在计算最终分包合同金额时，我方将考虑适当扣除与此类雇用有关的所有费用，或者作为债务从贵方收回。

敬上

抄送至：工程师

Letter 21

To sub-contractor that fails to comply with direction

Special delivery

Dear

I refer to the notice issued to you on the [insert date] under clause ____ of the conditions of sub-contract requiring compliance with our direction number [*insert number*] dated [*insert date*].

At the time of writing, you have not begun to comply with our direction. We are taking immediate steps to employ and pay others to comply with such direction. All costs incurred in connection with such employment will be taken into account in the calculation of the final sub-contract sum or will be recovered from you as a debt.

Yours faithfully

Copy: Engineer

信函 21

致未能遵照指示施工的分包商

快递

敬启者：

根据分包合同条款____，我方［插入日期］发布通知要求贵方遵照我方［插入日期］发出的，编号为［插入编号］的指示施工。

截止到目前，贵方还没有开始按照我方指示施工。我方正在采取紧急措施，雇用其他遵照该指示的人员施工并支付其费用。在计算最终分包合同金额时，我方将考虑适当扣除与此类雇用有关的所有费用，或者作为债务从贵方收回。

敬上

抄送至：工程师

Letter 22

To sub-contractor that wrongly confirms an oral direction

Dear

We are in receipt of your letter dated [*insert date*] in which you purport to confirm an oral direction which you allege was given by [*insert name*] on site/by telephone [*delete as appropriate*].

The issue of directions is governed by clause ____. We reject your confirmation, because there is no provision for oral directions nor for their confirmation by the sub-contractor.

Moreover, we deny that such a direction was given as you state or at all.

Yours faithfully

信函 22

致未能正确确认口头指示的分包商

敬启者：

我方收到贵方［插入日期］来函，贵方旨在确认一项贵方宣称由［插入名称］现场/电话［酌情删除］发出的口头指示。

条款____规定发布指示的相关问题。我方拒绝贵方的确认要求，因为合同中没有分包商发布口头指示的条款，也没有确认分包商口头指示的条款。

另外，我方否认发出贵方所宣称的口头指示或根本不存在口头指示。

敬上

Letter 23

To sub-contractor, inspection after failure of work

Dear

The [*describe work or materials*] are not in accordance with the sub-contract.

In accordance with clause ____ of the sub-contract and having due regard to the code of practice in Schedule 1, we enclose our directions for opening up for inspection/testing [*delete as appropriate*] which is reasonable in all the circumstances to establish to our reasonable satisfaction the likelihood or extent of any similar non-compliance.

Whatever the results may be, no adjustment will be taken into account in the calculation of the final sub-contract sum.

Yours faithfully

信函 23

致分包商，施工不符后的检验

敬启者：

[描述施工或材料] 不符合分包合同要求。

根据分包合同第____条款，并适当考虑附表1中的施工准则，我方附上指示要求打开施工项目检验/检测 [酌情删除]，在任何情况下，此做法均合理，旨在确定我方可以接受的类似违规情况或违规程度。

无论结果如何，在计算分包合同结算金额时，不会考虑任何费用调整。

敬上

Letter 24

To sub-contractor, after failure of work

Dear

We have today found that the [*describe the work, materials or goods*] are not in accordance with the sub-contract.

In accordance with clause ____, we require you to forthwith state in writing what action you propose to immediately take at no cost to us to establish that there is no similar failure in work already executed/materials or goods already supplied [*delete as appropriate*].

Be aware that if you do not respond within 5 days or if we are not satisfied with your proposals or if, because of safety considerations or statutory obligations, we are unable to wait for your proposals, we may issue directions requiring you at no cost to us to open up or test any work, materials or goods to establish whether there is a similar failure including making good thereafter.

Yours faithfully

信函 24

致分包商，在施工不符合要求之后

敬启者：

我方今天发现［描述施工、材料或货物］不符合分包合同要求。

根据条款____，我方要求贵方立即书面说明贵方打算采取何种措施以证明在已执行的施工中/已提供的材料或物品中［酌情删减］不存在类似不符合规定的情况，同时不能对我方造成任何费用损失。

请注意，如果贵方5天内没有答复，或我方不满意贵方提议，或者由于考虑到安全或法定义务问题，我方无法等待贵方提议，我方可以发布指示，要求贵方将施工项目、材料或货物打开检验以确定是否存在类似不合规情况，包括后期的修复工作在内，同时不能对我方造成任何费用损失。

敬上

Letter 25

To sub-contractor, fixing a new period for completion

Dear

We refer to your notice of delay dated [*insert date and if appropriate add*:] and the further information provided in your letter of the [*insert date*].

In accordance with clause ____ of the sub-contract, we hereby give you an extension of the period for completion of the sub-contract works of [*insert period*]. The revised period for completion of the sub-contract works is now [*insert period*].

The relevant sub-contact events taken into account are: [*list events in clause* ____ *and the periods granted in respect of each*].

We have attributed reduction in time to relevant sub-contract omissions as follows: [*list and include the extent of any reduction in extension of time*].

This is an interim/final [*delete a appropriate*] decision.

Yours faithfully

信函 25

致分包商，确定新的竣工期限

敬启者：

我方收到贵方［插入日期，酌情补充：］的延期通知以及贵方［插入日期］的函中所提供的进一步信息。

根据分包合同条款____，我方特此允许贵方延长分包合同工程竣工期限［插入时间期限］。现在分包合同工程完工的订正期限为［插入时间期限］。

考虑以下分包合同相关事项：［列出条款____中的事项及每一事项所允许的期限］。

我方认为减少时间的原因是相关分包合同任务的取消，减少时间如下：［列出并包括任何减少延长时间的程度］。

此为临时/最终［酌情删除］决定。

敬上

Letter 26

To sub-contractor, fixing a new period for completion after practical completion of the sub-contract works

Dear

In accordance with clause ____ and after reviewing all the evidence available to us [*add, if appropriate: 'including previous extensions of time'*], we

[*Add:*]

hereby extend the sub-contract period by [*insert additional period*]. The revised period for completion of the sub-contract works is now [*insert period*].

[Or:]

hereby shorten the sub-contract period by [*insert the reduction*] having regard to directions for relevant sub-contract omissions. The revised period for completion of the sub-contract works is now [*insert period*].

[*Or:*]

hereby confirm the period for completion previously fixed, namely [*insert period*].

Yours faithfully

信函 26

致分包商，分包合同工程实际完工后确定一个新的完工期限

敬启者：

根据条款____的规定，在审查提供给我方的所有证据后［酌情补充：包括前期的延长时间］，我方

［补充：］

特此将分包合同期限延长［插入增加的期限］。现在完成分包合同工程的订正期限为［插入期间］。

［或者：］

特此根据有关分包合同任务取消的指示，缩短分包合同期限［插入缩短期限］。现在完成分包合同工程的订正期限为［插入期限］。

［或者：］

特此确认以前确定的完工期限，即［插入期限］。

敬上

Letter 27

To sub-contractor，if claim for extension of time is not valid

Dear

We refer to your notice of delay dated [*insert date and if appropriate add*:] and the further information provided in your letter dated [*insert date*].

On the basis of the documents you have presented to us，we see no ground for any extension of time. We shall be pleased to consider any further submissions if they are presented in the proper form and in accordance with the terms of the sub-contract.

Yours faithfully

信函 27

致分包商，如延期要求无效

敬启者：

我方收到贵方［插入日期并酌情补充：］的延期通知以及贵方在［插入日期］的函中提供的进一步信息。

根据贵方向我方提交的文件，我方认为贵方无理由延期。如以正确形式，并依据分包合同条款要求提交文件，我方将愿意考虑任何进一步提交的建议。

敬上

Letter 28

To sub-contractor, if sub-contract works not complete within the period for completion

Dear

In accordance with clause ____ of the sub-contract, we hereby give notice that the sub-contract [*insert nature of works*] works were not completed within the period for completion/revised period for completion [*delete as appropriate*] ending on [*insert date which should not be unreasonably earlier than the date of this letter*].

You are in breach of the sub-contract and we draw your attention to clause ____ and our right to recover from you the amount of any direct loss and/or expense which we have suffered or incurred as a result of your failure to complete in due time. We reserve all our rights and remedies.

Yours faithfully

信函 28

致分包商，如未能在完工期限内完成分包合同

敬启者：

根据分包合同条款____，我方特此通知分包合同［插入工程性质］工程未在完工期/修订完工期［酌情删除］内完成，应于［插入日期，应早于本函日期］结束工程。

贵方违反分包合同，我方提请贵方注意条款____，我方有权向贵方追索因贵方未能在规定时间内完成工程而造成的任何直接损失和/或费用。我方保留一切权利及补救措施。

敬上

Letter 29

To sub-contractor，applying for payment of loss and/or expense

Dear

We hereby give notice and make application under clause ____ of the sub-contract as follows：

We have been caused direct loss and/or expense because the regular progress of the Works has been materially affected by [*describe*].

Particulars of the calculation of such direct loss and/or expense are enclosed and we should be pleased to have your agreement by close of business on [*insert date*] to the amount of [*insert amount*].

Yours faithfully

信函 29

致分包商，申请支付损失和/或费用赔偿

敬启者：

根据分包合同条款____项，我方特此通知并作如下申请：

我方已经遭受直接损失和/或费用开支，原因是［描述］对工程正常进展造成实质性影响。

附上此类直接损失和/或费用开支的计算详情，我方希望贵方在［插入日期］下班前同意我方提出的损失金额［插入金额］。

敬上

Letter 30

To sub-contractor，giving notice of an interim payment

Dear

This is a written notice specifying the amount of interim payment which is proposed to be made，namely：[*insert amount*]. The amount is calculated on the following basis：[*insert the way in which the amount is calculated. If it is by reference to a certificate*, *so state*].

Yours faithfully

信函 30

致分包商，发出临时付款通知

敬启者：

该书面通知详述了拟支付的临时付款金额，即：[插入金额]。金额计算依据如下：[插入金额计算方式。如果是通过参考凭证进行计算，请说明]。

敬上

Letter 31

To sub-contractor, if sub-contractor has correctly sent 7 day notice of intention to suspend performance of obligations

Dear

We are in receipt of your letter of the [*insert date*] which you have sent as a notice in accordance with clause ____ of the sub-contract.

We regret that you have felt it necessary to send such a notice, but we are pleased to enclose our cheque in the sum of [*insert amount*]. We hope you will accept our apologies for this oversight.

Yours faithfully

信函 31

致分包商，如果分包商已正确发出了中止履行义务 7 天的意向通知

敬启者：

收悉贵方于［插入日期］依据分包合同条款____所发的通知。

贵方认为有必要发出此类通知，对此我方深感遗憾。但我方很愿意附上金额为［插入金额］的支票。望贵方接受我方对此疏忽的歉意。

敬上

Letter 32

To sub-contractor，requesting documents for calculation of the final sub-contract sum

Dear

In accordance with clause ____ of the sub-contract，we should be pleased to receive all documents necessary for the purpose of calculating the final sub-contract sum.

Yours faithfully

信函 32

致分包商，索要用来计算分包合同结算金额的文件

敬启者：

根据分包合同条款____，我方希望收到用来计算分包合同结算金额所需的全部文件。

敬上

Letter 33

To sub-contractor that has failed to submit documents for the calculation of the final sub-contract sum (a)

Dear

We note that you have failed to submit the documents required under clause ____ for the purpose of calculating the final sub-contract sum.

It is now more than two months since the practical completion of the sub-contract works and it is my duty under clause ____ of the sub-contract to prepare a statement of all adjustments to the final sub-contract sum as we can make based on the information we have. A copy of such statement is enclosed.

Yours faithfully

信函 33

致分包商，分包商未提交用于计算分包合同结算金额的文件（a）

敬启者：

我方注意到贵方未提交合同条款____要求的用以计算分包合同结算金额的文件。

现在分包合同工程实际完工已超两月。分包合同第____条要求我方负责准备一份所有分包合同结算金额调整的说明，我方可以依据我方掌握的信息作出此说明。附上此说明的副本。

敬上

Letter 34

To sub-contractor that has failed to submit documents for the calculation of the final sub-contract sum (b)

Dear

We note that you have failed to submit the documents required for the purpose of calculating the final sub-contract sum.

The time period allotted in the sub-contract for submission of the documents has now expired and we will now proceed to calculate the final sub-contract sum using the information in our possession.

Your failure to submit the documents is a clear breach of the sub-contract. Although we will calculate the sub-contract sum as fairly as possible, you cannot expect to profit from your breach and, in the absence of appropriate evidence, we shall certainly not make any financial assumptions in your favour.

Yours faithfully

信函 34

致分包商，分包商未提交用于计算分包合同结算金额的文件（b）

敬启者：

我方注意到贵方没有提交用于计算分包合同结算金额所需的文件。

分包合同指定提交该文件的截止日期已过，我方现在将根据已有信息计算分包合同结算金额。

贵方未能提交文件明显违反了分包合同。我方将尽可能公平地计算分包合同金额，但贵方不要指望从该违约中获利。由于缺少适当的凭证，我方无法作出对贵方有利的财务假设。

敬上

Letter 35
To employer, giving notice of the named sub-contractor's default

Dear

We hereby notify you that, in our opinion, [*insert name*] has made default in [*insert nature of default*] being a matter referred to in clause [*insert clause number as appropriate to the particular sub-contract giving grounds for termination*] of the sub-contract [*insert name of sub-contract form*]. A copy of the clause is attached for your convenience.

Because [*insert name*] is a named sub-contractor under supplemental provision, Schedule ____, paragraph ____, we request your consent under paragraph ____ to our intention to terminate the sub-contractor's employment.

Yours faithfully

信函 35
致雇主，通知雇主提名分包商的违约行为

敬启者：

特此通知贵方，我方认为［插入名称］的行为［插入违约性质］已违反分包合同［插入分包合同格式名称］中的条款规定［插入具体说明终止原因的分包合同条款序号］。为方便起见，随函附上该条款的副本。

由于［插入名称］是附表____，第____段补充条款中的提名分包商，因此，根据段落____中内容，要求贵方批准我方与该分包商终止雇用关系。

敬上

Letter 36

To sub-contractor，giving notice of default before termination

Special delivery

Dear

We hereby give you notice under clause ____ of the sub-contract that you are in default in the following respect：

[*Insert details of the default with dates if appropriate.*]

If you continue the default for 10 days after receipt of this notice or if you at any time repeat such default，whether previously repeated or not，we may within 10 days of such continuance or within a reasonable time of such repetition terminate your employment under this sub-contract.

Yours faithfully

信函 36

致分包商，在终止雇用关系前通知违约行为

快递

敬启者：

根据分包合同条款____，特此通知贵方在以下方面存在违约行为：

[插入违约细节，酌情附加日期。]

如果贵方在收到通知后 10 日内仍继续违约或任何时间重复此违约行为，无论此前是否重复，根据分包合同规定，我方将于贵方继续违约的 10 日内或重复违约的合理时间内与贵方终止雇用关系。

敬上

Letter 37

To sub-contractor, giving notice before termination

Special delivery

Dear

We hereby give you notice under clause ____ of the sub-contract that the following ground has arisen:

[*Insert details of the ground, which should be contained in clause ____, with dates if appropriate.*]

If the ground is still in existence for 10 days after receipt of this notice or if it arises again at any time thereafter we may terminate this sub-contract.

Yours faithfully

信函 37

致分包商，终止分包合同前通知分包商

快递

敬启者：

根据分包合同条款____，特此通知贵方出现下述情况：

[插入情况的细节信息，这些信息应包含于条款____中，酌情添加日期。]

如在收到本通知后10日内仍存在该情况，或在此后的任何时间再次出现，我方可终止本分包合同。

敬上

Letter 38

To engineer，if termination of named person's employment possible

Dear

In accordance with ____ of the conditions of contract，we have to advise you that the following events are likely to lead to the termination of [*insert name*] 's employment：

[*Describe the events*.]

We should be pleased to receive your instructions.

Yours faithfully

信函 38

致工程师，如果有可能解除与提名人员的劳动合同

敬启者：

根据合同条款____，我方必须提醒贵方下述情况可能导致我方与［插入姓名］终止劳动合同。

［描述事件。］

期待收到贵方指示。

敬上

Letter 39

To sub-contractor，terminating employment after default notice

Special delivery

Dear

We refer to the default notice sent to you on the [*insert date*] .

Take this as notice that，in accordance with clause ____，we hereby terminate your employment under this sub-contract without prejudice to any other rights or remedies which we may possess.

The rights and duties of the parties are governed by clause ____ . No temporary buildings，plant，tools，equipment，goods or materials shall be removed from site until and if we so direct. Take note that any other provisions which require any further payments or release of retention cease to apply.

Yours faithfully

信函 39

致分包商，告知违约行为后终止雇用合同（a）

快递

敬启者：

谨提及我方于［插入日期］发送给贵方的违约通知。

根据条款____，我方特此通知与贵方解除本分包合同下的雇用关系，这不影响我方拥有的任何其他权利和补救措施。

条款____明确规定了双方的权利和义务。在我方作出指示并同意之前，不得从施工现场移除任何临时建筑物、厂房、工具、设备、物品或材料。请注意，所有其他要求另外付款或退回保留金的合同条款停止适用。

敬上

Letter 40

To sub-contractor, terminating the sub-contract after notice
Special delivery

Dear

We refer to the clause ____ notice sent to you on the [*insert date*] *n*.

Take this as notice that, in accordance with clause ____, we hereby determine this sub-contract without prejudice to any other rights or remedies which we may possess.

We shall give directions under clause ____ in due course.

Yours faithfully

信函 40

致分包商，通知后终止分包合同（b）
快递

敬启者：

谨提及我方于［插入日期］根据条款____发出的通知。

根据条款____，我方特此通知终止此分包合同，这不影响我方拥有的任何其他权利或补救措施。

我方届时将根据条款____发布指示。

敬上

Letter 41

To sub-contractor, termining the sub-contract without prior notice

Special delivery

Dear

The following ground for termination has arisen:

[*Insert details of the grounds which should be contained in clause* ____ *other than clauses* ____ .]

Therefore, take this as notice that, in accordance with clause ____, we hereby determine this sub-contract without prejudice to any other rights or remedies which we may possess.

We shall give directions under clause ____ in due course.

Yours faithfully

信函 41

致分包商，在未预先通知的情况下终止分包合同

快递

敬启者：

终止分包合同的原因如下：

[插入条款____中应包含的理由的细节，但不包括条款____。]

因此，我方根据条款____特此通知终止本分包合同，这不影响我方拥有的其他任何权利和补救措施。

我方届时将根据条款____发布指示。

敬上

Letter 42

To sub-contractor, terminating employment after termination of the main contract

Special delivery

Dear

In accordance with clause ____ of the sub-contract, we have to inform you that our employment under the main contract was terminated on the [*insert date, bearing in mind that this letter must be sent immediately the main contract termination takes place*]. The clause provides that your employment under this sub-contract must thereupon terminate.

The rights and duties of the parties are governed by clause ____. Take note that any other provisions which require any further payments or release of retention cease to apply. All our other rights and remedies are preserved.

Yours faithfully

信函 42

致分包商，终止主合同后解除雇用关系

快递

敬启者：

根据分包合同条款____，必须通知贵方，我方在主合同下的雇用关系已于［插入日期，切记，此函必须在主合同终止后立即发送］终止。该条款规定本分包合同下的贵方雇用关系必须随之终止。

条款____规定了双方的权利与义务。请注意，所有其他要求另外付款或退回保留金的合同条款停止适用。我方保留其他所有权利和补救措施。

敬上

第十章　索赔与仲裁信函

第一节　索　赔　概　述

一、索赔含义

索赔的本意是主张自身权益，当一方受到损失时向另一方提出补偿要求。它是一种正当权利的要求，并不意味着对过错方的惩罚，所以其基调是温和的。无论是一方违约使另一方蒙受损失情况下的索赔，还是双方都没有违约而因自然灾害引起的索赔，都是一种正当权利的要求。索赔不是法律上的概念，在建设工程施工合同中，索赔是作为一个章节出现的，由此可见索赔是建设工程管理的重要手段和必要环节，它既包括承包商向雇主的索赔，也包括雇主向承包商的索赔。索赔要遵守一定的索赔程序，按照索赔程序提出索赔要求，提交索赔证据。只有当索赔没有得到被索赔方的认可或接受，索赔方仍然坚持维护自己的权益时，索赔才可能转向法律途径，即通过仲裁或诉讼解决索而未决的问题。

本章的索赔信函主要侧重招致损失和费用方面的信函写作。

二、合同文件引起的索赔

（一）合同文件组成引起的索赔

建设工程施工合同是一个由合同协议书、通用条款、特殊条款等各种法律文件组成的文件综合体，有些合同文件是在中标后通过讨论修改补充的，如果没有专业的合同管理，很可能会造成合同文件的前后矛盾和歧义。因此，在 FIDIC 合同文件的组织和管理过程中，明确各种合同文件的效力次序是非常重要的。如果因疏忽未加以明确，极有可能造成索赔事件的发生。

（二）合同缺陷引起的索赔

合同缺陷是指合同条款规定不严谨甚至前后矛盾，合同中有遗漏或者错误。它不仅包括条款中的缺陷，也包括技术规程和图纸上的缺陷。工程师有权对缺陷作出解释，但如果承包商执行工程师的解释后引起了成本增加或工期延误，则有权提出索赔。

三、不可抗力和不可预见因素引起的索赔

（一）不可抗力的自然因素引起的索赔

不可抗力的自然因素是指飓风、地震、海啸等自然灾害。FIDIC 合同条件规定，由这类自然灾害引起的损失应由雇主承担，但是 FIDIC 也指出，承包商在这种情况下应积极采取措施，减小损失。

（二）不可抗力的社会因素引起的索赔

不可抗力的社会因素是指发生战争、核装置的污染、暴乱、承包商和其分包商以外人员的动乱和骚扰等风险。这些风险一般应由雇主承担，承包商应得到损害前已完成的永久工程的付款和合理利润以及一些修复费用和重建费用。

（三）不可预见的外界条件引起的索赔

不可预见的外界条件是指有经验的承包商在招标阶段根据招标文件提供的资料和现场勘察情况，仍无法合理预见到的外界条件，如地下水、地质断层、溶洞等，但其中不包括气候条件（异常恶劣天气条件除外）。遇到此类条件，承包商受到损失或增加额外支出，经工程师确认，承包商可获得经济补偿和工期顺延。如工程师认为承包商在提交投标书前根据已知的现场条件、地质勘探资料应能预见到的情况，承包商在投标时理应予以考虑，可不同意索赔，风险由承包商承担。

（四）施工中遇到地下文物或构筑物引起的索赔

在挖方工程中，如发现图纸中未注明的文物（不管是否有价值）或人工障碍（如公共设施、隧道旧建筑物等），承包商应立即报告工程师，进行现场检查，共同讨论处理方案。如果新施工方案导致工程费用增加，如原计划的机械开挖改为人工开挖等，承包商有权提出费用索赔和工期索赔。

四、雇主原因引起的索赔

（一）拖延提供现场及通道引起的索赔

因施工现场的搬迁工作进展不顺利等原因，雇主没能如期向承包商移交合格的、可以直接进行施工的现场，会导致承包商提出误工的费用索赔和工期索赔。

（二）拖延支付工程款引起的索赔

合同中均有支付工程款的时间限制，如果雇主不能按时支付工程进度款，承包商可按合同规定向雇主索赔利息。严重拖欠工程款而使承包商资金周转困难时，承包商除向雇主提出索赔要求外，还有权放慢施工进度，甚至可以因雇主违约而解除合同。

（三）雇主提前占用部分永久工程引起的索赔

在整体工程接收前，雇主要求对部分单项工程提前使用，如果合同未规定可提前占用部分工程，则提前使用永久工程的单项工程或部分工程所造成的后果，由雇主承担；另一方面，提前占用工程若影响承包商的后续工程施工，或影响承包商的施工组织计划，增加施工困难，承包商有权提出索赔。

（四）雇主要求提交施工计划引起的索赔

由于雇主改变原合同规定，或者改变了部分工程的施工内容而必须延长工期，从而导致成本增加，承包商可以要求赔偿赶工措施费用，例如加班工资、新增设备租赁费和使用费、增加的管理费用、分包的额外成本等。

五、工程师原因引起的索赔

（一）延误提供图纸或拖延审批图纸引起的索赔

如工程师延误向承包商提供施工图纸，或者拖延审批承包商负责设计的施工图纸，因此而使施工进度受到影响，承包商可以索赔工期，还可以对延误导致的损失要求经济补偿。

（二）违反程序性规定引起的索赔

FIDIC、AIA等国际工程标准合同文本都对施工现场管理有严格的程序性规定和时间的限制性规定。工程师作为雇主人员，在施工现场代理雇主对施工过程进行管理，如果工程师不尽职，延误了对工程施工文件的接收或回复，造成损害的发生，承包商可以提出索赔。

（三）工程质量要求过高引起的索赔

在施工过程中，工程师有时会不认可承包商提供的某种材料，而迫使其使用比合同文件规定的标准更高的材料，或者提出更高的工艺要求，则承包商可以要求对其损失进行补偿或重新核定单价。

（四）对承包商的施工进行不合理干预引起的索赔

在施工管理过程中，工程师对承包商的施工顺序及施工方法进行不合理的干预，甚至超出合同范围，下达指令要求承包商执行，则承包商可以就这种干预所引起的费用增加和工期延长提出索赔。

（五）暂停施工引起的索赔

项目实施过程中，工程师有权根据承包商违约或破坏合同的情况，或者因现场气候条件不利于施工以及为了工程的合理进行而有必要停工时，下达暂停施工的指令。如果这种暂停施工的命令并非因承包商的责任或原因所引起的，则承包商有权要求工期赔偿，同时可以就其停工损失获得合理的额外费用补偿。

第二节　仲　裁　概　述

一、国际工程仲裁的含义

仲裁是双方当事人根据书面仲裁协议，将所约定的争议事项提交约定的仲裁机构进行审理，并由该机构作出具有约束力的仲裁裁决的一种争议解决方式。如果国际工程合同规定采取司法程序解决争议，或者在合同中未提及争议解决的方式且双方未达成一致，那么，此类争议就需要通过司法程序，即通过国际工程诉讼的方式来解决。争议的任何一方都有权向有管辖权的法院起诉。

国际上知名的仲裁机构包括：国际商会仲裁院（The International Court of Arbitration of International Chamber of Commerce，ICCCA）、斯德哥尔摩商会仲裁院（The Arbitration Institute of the Stock-holm Chamber of Commerce，SCCCA）、中国国际经济贸易仲裁委员会（China International Economic and Trade Arbitration Commission，CIETAC，又称中国国际商会仲裁院）、伦敦国际仲裁院（The London Court of International Arbitration，LCIA）等。

国际工程仲裁地点一般是当事人合意的选择。在没有特殊约定时，通常将被选定的常设仲裁机构所在地作为仲裁地点。国际工程仲裁中，仲裁地点是一个至关重要的因素。雇主一般要求在项目所在国的仲裁机构进行仲裁，而承包商则希望在承包商总部所在国的仲裁机构进行仲裁，常见的妥协方案是仲裁地点设在第三国或被申请人的国家。在仲裁中，双方可以约定仲裁语言，一般采用的语言与合同执行所用语言一致，仲裁常用语言为英语。

一裁终局是国际仲裁制度的一种重要的法律特性和优势。绝大多数国家的仲裁立法和司法实践以及仲裁实务都认可，仲裁机构作出的裁决具有终局效力，对合同双方具有法律约束力，任何一方不得上诉或申诉。对于国际仲裁裁决，1958 年的《承认及执行外国仲裁裁决公约》作出了明确规定。截止到 2007 年 9 月，该公约目前的签字国有 142 个，我国于 1987 年参加了该公约。这一公约保证了国际仲裁裁决的可执行性。

本章主要介绍发出提交争议仲裁意向书、提名仲裁员等与仲裁相关的信函。

二、FIDIC“新红皮书”关于仲裁条款的规定

除非通过友好解决，否则如果争议评判委员会有关争端的决定（如有时）未能成为最终决定并具有约束力，那么此类争端应由国际仲裁机构最终裁决。

同时FIDIC赋予了仲裁人一定的权利，即仲裁人应有权公开、审查和修改工程师的任何证书的签发、决定、指示、意见或估价，以及任何争议评判委员会有关争端事宜的决定。无论如何，工程师都不会失去作为证人以及向仲裁人提供任何与争端有关的证据的资格。

FIDIC合同条件下的仲裁原则：

(1) 合同双方的任一方在上述仲裁人的仲裁过程中均不受以前为取得争议评判委员会的决定而提供的证据或论据或其不满意通知中提出的不满理由的限制。在仲裁过程中，可将争议评判委员会的决定作为一项证据。

(2) 工程竣工之前或之后均可开始仲裁。但在工程进行过程中，合同双方、工程师以及争议评判委员会的各自义务不得因任何仲裁正在进行而改变。

第三节 索 赔 信 函

Letter 1

To engineer, applying for payment of loss and/or expense (a)

Dear

We hereby make application under clause ____ of the conditions of contract as follows:

We have incurred/are likely to incur [*delete as appropriate*] direct loss and/or expense and financing charges in the execution of this contract for which we will not be reimbursed by a payment under any other provision in this contract, because the regular progress of the Works has been/is likely to be [*delete as appropriate*] materially affected by [*describe and unless referring to deferment of possession, add*:] being a relevant matter in clause [*insert clause number*].

Yours faithfully

信函 1

致工程师，申请支付损失及/或费用 (a)

敬启者：

依据合同条款____，我方特此提出以下申请：

在执行本合同过程中，我方已承担/有可能承担［酌情删除］直接损失及/或费用以及融资费用，根据本合同的任何其他条款，我方不会得到这些费用的支付补偿。原因是工程的正常进度已经/有可能［酌情删除］受到相关事项［描述并且除非提及所有权押后问题，补充：］的实质影响，该事项包括在条款［插入条款编号］中。

敬上

Letter 2

To engineer, applying for payment of loss and/or expense under the supplemental provisions (b)

Special delivery

Dear

We are entitled to an amount in respect of loss and/or expense (including financing charges) to be added to the contract sum in accordance with clause ____ of the conditions of contract. Our application for payment under clause ____ is attached. Therefore in accordance with supplemental provision ____, we submit our estimate of such loss and/or expense, incurred in the period immediately preceding that for which such application is made, which we require to be added to the contract sum. We shall continue to submit estimates in accordance with paragraph ____ for so long as we continue to incur direct loss and/or expense.

May we remind you that within 21 days of receipt of this estimate, you must give us written notice either that you accept our estimate, or that you wish to negotiate and in default of agreement that you wish to refer the issue as a dispute to adjudication, or clause ____ will apply.

Yours faithfully

信函 2

致工程师，根据补充条款申请支付损失及/或费用（b）

快递

敬启者：

依照合同条款____，我方有权就损失及/或费用（包括融资费用）追加合同金额，我方已随函附上条款____中要求的支付申请。因此，依据附加条款____的规定，我方提交相关损失及/或费用的估算金额，该笔费用产生后我方立即提出了支付申请，我方要求将其追加到合同总额中。只要我方继续承担直接损失及/或费用，我方将继续按照合同第____段规定提交估算金额。

望贵方注意，在收到估算值起 21 天内，务必给予我方书面通知，接受我方估算金额，或希望与我方协商且不同意将争议事项提交仲裁，或条款____将适用。

敬上

Letter 3

To engineer, applying for payment of loss and/or expense (c)

Dear

We hereby give notice that regular progress of the Works has been materially disrupted/prolonged [*delete as appropriate*] by matters for which the employer or you as the employer's agent are responsible. In our view, such matters exceed the situation contemplated by clause ____ of the conditions of contract. The matters are [describe].

Although the contract appears to have no machinery for dealing with claims of this nature, we are advised that we may bring an action at common law. We believe that the matters are capable of easy resolution by the employer or by you with proper authority.

We should be content to proceed on this basis and we should be pleased to hear whether the employer is in agreement. If the employer is not prepared to deal with us on this basis or if we do not hear from you by [*insert date*], we shall formulate our claim for damages at common law.

Yours faithfully

Copy: Employer

信函 3

致工程师，申请支付损失及/或费用（c）

敬启者：

我方特此通知由雇主或作为雇主代理的贵方所负责的事项已对工程的正常进度造成实质性的干扰/延误［酌情删除］。我方认为此类事项超出合同条款所规定的情况。此类事项为［描述情况］。

虽然合同中似乎没有解决此类索赔的规定，但是我方获悉我方可根据普通法提起诉讼。我方相信这些事项能够由雇主轻易解决或贵方有权轻易解决。

在此基础上，我方愿意继续施工，期待雇主同意我方建议。如果雇主不准备在此基础上接受我方建议，或我方未能于［插入日期］前收到贵方回复，我放将依照普通法申索损失赔偿。

敬上

抄送至：雇主

Letter 4

To engineer, applying for payment of expense (d)

Dear

We hereby give notice under clause ____ of the conditions of contract as follows:

The regular progress of the Works has been/is likely to be [*delete as appropriate*] disrupted/prolonged [*delete as appropriate*] due to [*describe*] . In consequence of such disruption/prolongation [*delete as appropriate*] we have properly and directly incurred/ we will properly and directly incur [*delete as appropriate*] expense in performing the contract which we would not otherwise have incurred and which is beyond that otherwise provided for in or reasonably to be contemplated by the contract.

We are/we expect to be [*delete as appropriate*] entitled to an increase in the contract sum under clause ____ .

Yours faithfully

信函 4

致工程师，申请支付费用（d）

敬启者：

我方根据合同条款____，特此通知如下：

由于［描述］，工程的正常进度已经/有可能［酌情删除］中断/延长［酌情删除］。因为该中断/延长，在执行合同过程中，我方已直接承担/将直接承担一定［酌情删除］费用，我方本不该承担该费用，且该费用不包括合同另有规定或合理预期的费用。

根据条款____，我方有/期望有［酌情删除］权增加合同金额。

敬上

Letter 5

To engineer, giving further details of loss and/or expense

Dear

Thank you for your letter of the [*insert date*] requesting further information in support of our application for loss and/or expense which was submitted to you on [*insert date*].

We enclose a copy of the programme in precedence diagram/network analysis [*delete as appropriate*] form, which we have marked up to show the circumstances in some detail. It can be seen that [*describe the circumstances in some detail, giving dates and times, numbers of operatives and names of key members of staff involved*].

Also enclosed is the following copy correspondence, extracts from the site diary and site minutes:

[*List with dates. Do not include the following paragraph when using DB:*]

We should be pleased if you would inform us if there are any particular points on which you require more information before you are able to form an opinion as required by clause ____ of the conditions of contract.

Yours faithfully

信函 5

致工程师，提供损失和/或费用的详情

敬启者：

奉读贵方［插入日期］来函，函中要求我方提供详细信息证明我方于［插入日期］所呈递的损失和/或费用申请。

随函附寄项目进展顺序图/网状分析图［酌情删除］副本，并已在某些细节上进行标明以详细阐述情况。如此可见［用一些细节阐述该情况，提供日期和时间，作业人员的数目和主要成员的姓名］。

另随函附上以下通信副本，其内容摘自工地日志和工地会议记录：

［日期列表。当采用 DB 模式时删除下段内容：］

在贵方根据合约条款____要求形成意见之前，如果有任何需要提供更多信息的特定问题，请贵方通知我方，我方将欣然提供。

敬上

Letter 6
To engineer or quantity surveyor, enclosing details of loss and/or expense

Dear

Thank you for your letter of the [*insert date*] requesting details of loss and/or expense in respect of the matters notified in our letter of [*insert date*] .

We enclose the details together with supporting documentation. We should be pleased if you would proceed with the ascertainment of loss and/or expense as required by clause ____ . Please inform us immediately if you require any further information.

Yours faithfully

信函 6
致工程师或估算师，随函附寄损失和/或费用详情

敬启者：

感谢贵方［插入日期］来函，要求我方就［插入日期］函中通知的事项提供损失及/或费用详情。

随函附寄详情与证明文件。希望贵方能依照条款____对损失及/或费用进行确认。如果贵方需要任何进一步信息，请立即通知我方。

敬上

Letter 7

To quantity surveyor，providing information for calculation of expense

Dear

We have pleasure in enclosing the documents listed below. They contain full details of all expenses incurred and evidence that the expenses directly result from the occurrence of one of the events described in clause ____.

We look forward to your decision in accordance with clause ____ within 28 days of receipt of this letter.

[*List documents and information.*]

Yours faithfully

信函 7

致估算师，提供费用计算所需信息

敬启者：

我方随函附寄以下文件。文件内容包括所有已发生费用的详情及合同条款____所述事项直接导致的费用证明。

希望贵方能于收函起 28 天之内根据条款____作出决定。

[列举文件和信息。]

敬上

Letter 8

To engineer, if ascertainment delayed (a)

Dear

We refer to our notice of the [*insert date*] submitted under the provisions of clause ____ of the conditions of contract and our letters of the [*insert dates*] enclosing the further information you required in order to form an opinion and carry out ascertainment of the amount of loss and/or expense.

[*Insert number*] weeks have elapsed since we last submitted such information to you and, during that period, neither you nor the quantity surveyor have requested further information or details.

Clause ____ imposes a clear duty on you to certify sums ascertained in whole or in part in the next interim certificate following ascertainment.

Clause ____ states that 'as soon as' you are of the opinion that the Works are affected, you will 'from time to time thereafter' ascertain the amount of loss and/or expense. A similar provision is contained in clause ____ .

As soon as you have formed your opinion, you must begin the process of ascertainment.

As soon as any sum has been ascertained, you must include the amount in the next interim certificate. Please inform us by [*insert date*] the amount which you have ascertained and intend to include in the next certificate.

Yours faithfully

信函 8

致工程师，如确认事项被延误（a）

敬启者：

兹谈及我方［插入日期］依照合同条款____规定所呈递的通知及［插入日期］信函，函中附有贵方形成意见和确认损失及/或费用金额所需的进一步信息。

自我方提交以上信息起已过［插入数字］周，在此期间，贵方和估算师均未要求我方提供进一步信息和详情。

条款____规定贵方有义务核证下一个临时证书中确认的全部或部分款项。

条款____规定贵方"一旦"认为工程受到影响，贵方要"自此经常"确认损失及/或费用金额。条款____中包含类似规定。

贵方一旦形成意见，就必须启动确认过程。任何费用金额一旦确认，贵方必须将其纳入下一临时证书中。请于［插入日期］前将贵方已确认并打算纳入下一证书的金额告知我方。

敬上

Letter 9
To engineer, if ascertainment delayed (b)

Dear

We refer to our notice of the [*insert date*] submitted under the provisions of clause ____ of the conditions of contract and our letters of the [*insert dates*] enclosing the further information you required to form an opinion and carry out ascertainment of the amount of loss and/or expense.

[*Insert number*] weeks have elapsed since we last submitted such information to you and, during that period, neither you nor the quantity surveyor have requested further information or details.

Clause ____ imposes a duty on you to certify ascertainment under clause ____ in your certification of interim payments 'to the extent that it has been ascertained'.

Please inform us by [*insert date*] the amount which you have ascertained and intend to include in the next certificate.

Yours faithfully

信函 9
致工程师，如果确认事项被延误（b）

敬启者：

兹谈及我方［插入日期］依照合同条款____规定所呈递的通知及［插入日期］信函，函中附有贵方形成意见和确认损失及/或费用金额所需的进一步信息。

自我方提交以上信息起已过［插入数字］周，在此期间，贵方和估算师均未要求提供进一步信息和详情。

条款____规定贵方有义务按照条款____要求在临时付款证书中证明认定金额已“全额认定”。

请于［插入日期］前将贵方已认定并打算纳入下一证书的金额告知我方。

敬上

Letter 10

To engineer, if ascertainment too small

Dear

We refer to our notice dated [*insert date*] in respect of loss and/or expense. Your letter of the [*insert date*] notifying us of the amount ascertained appears to take little account of the very full supporting information submitted by us on [*insert date or dates*].

Unless we hear from you by [*insert date*] that you will amend your ascertainment to take account of the information we have supplied, a dispute will have arisen which we will refer to adjudication in due course.

Yours faithfully

信函 10

致工程师，如果确认金额太少

敬启者：

兹谈及我方［插入日期］涉及损失和/或费用的通知。贵方［插入日期］函告我方的确认金额似乎并未考虑我方［插入日期］提交的所有证明信息。

除非在［插入日期］前收到贵方来函，说明贵方将根据我方提供的信息修改认定金额，否则我方将适时采取法律措施裁决由此产生的争议。

敬上

Letter 11

To employer，regarding a common law claim

Dear

We draw your attention to [*describe the circumstances giving rise to the claim with dates*]. These circumstances entitle us to a claim for damages against you.

The contract makes no provision for such a claim in such circumstances and we intend to take immediate steps to recover under the contractual dispute resolution procedures.

Yours faithfully

Copy：Engineer

信函 11

致雇主，关于普通法索赔

敬启者：

提请贵方注意［陈述引起索赔的情况，日期］。我方有权就此类情况向贵方要求损失赔偿。

在这种情况下，合同未对该索赔做出规定，我方打算依据合同争议解决程序立即采取措施提出索赔。

敬上

抄送至：工程师

Letter 12
To employer，regarding a common law claim

without prejudice

Dear

We refer to the claim we have against you at common law arising from the circumstances notified to you in our letter of the [*insert date*].

If you are prepared to meet us to discuss our claim with a view to reaching a reasonable settlement of the matter，we will take no immediate legal steps.

Please inform us by [*insert date*] if you agree to this suggestion and let us know a date when such a meeting could take place.

Yours faithfully

Copy：Engineer

信函 12
致雇主，关于普通法索赔

无损法定利益

敬启者：

兹谈及就普通法向贵方索赔事宜，我方已于［插入日期］函告贵方导致索赔之情况。

如果贵方准备与我方会晤讨论索赔事宜，以期达成这一问题的合理解决，我方将不会立即采取法律措施。

如果贵方同意该建议，请于［插入日期］前通知我方，并告知我方会谈日期。

敬上

抄送至：工程师

第四节　仲　裁　信　函

Letter 1

To employer, giving notice of intention to refer a dispute to adjudication

Special delivery

Dear

Under the provisions of clause ____ and the Scheme for Construction Contracts (England and Wales) Regulations 1998 we intend to refer the following dispute or difference to adjudication: [*insert a description of the dispute including where and when it arose*]. We will be requesting the adjudicator to [*insert the nature of the redress sought e.g., 'order immediate payment of the outstanding amount of X or such sum as the adjudicator decides is due'*].

For the record, the names and addresses of the parties to the contract are as follows: [*set out the names and addresses which have been specified for the giving of notices*].

[*Either, if the adjudicator is named in the contract, add:*]

The adjudicator will be [*insert name*] as specified in the contract particulars/abstract of particulars [*delete as appropriate*].

[*Or, if the named adjudicator is unable to act or if none is named, add:*]

We are applying to [*insert the name of the nominating body*] for the nomination of an adjudicator.

Yours faithfully

Copy: Adjudicator/nominating body [*delete as appropriate*]

信函 1

致雇主，发出提交争议仲裁意向书

快递

敬启者：

根据条款____和《建筑合同格式（英格兰与威尔士）条例 1998》的规定，我方打算将下述争议或分歧提交裁决：[补充争议描述，包括发生的地点和时间]。我方将向裁决人发出请求［补充索求赔偿的性质，如“要求立即支付债务总额 X 或裁决者判决的应付金额”]。

供记录在案，合同双方的名称和地址如下：［列出所发通知中具体说明的名称和地点]。

［或者，如果仲裁员没有在合同中指定，请补充：]

仲裁员将是［插入名字]，此为合同细节/细节摘要［酌情删除］指定的。

［或者，如果指定的仲裁员无法执行仲裁，或者如果没有指定仲裁员，则补充：]

我方申请指定［插入提名机构的名称］为仲裁人。

敬上

抄送至：仲裁人/指定机构［酌情删除]

Letter 2

To nominating body，requesting nomination of an adjudicator

Special delivery

Dear

We enclose a notice of intention to refer a dispute and/or difference under the contract to adjudication. We have today served this notice and covering letter on the other party to the contract：[*insert the name of the employer*]. The contract was executed on [*insert the name of the contract as it appears on the cover*] terms. You are the selected nominator. Therefore，in accordance with clause 9.2 and paragraph 2 of the Scheme for Construction Contracts（England and Wales）Regulations 1998，we hereby make application to you to select a person to act as Adjudicator. A copy of the completed application form and the Referring Party's cheque in the sum of [*insert the amount*] is enclosed.

[*If appropriate，add*：]

We should be grateful if you would not nominate any of the following persons：[*list any adjudicators on the panel of the nominating body whom you do not wish to be nominated in this instance and give brief reasons*].

Yours faithfully

Copy：Employer

信函 2

致指定机构，请求指定一名仲裁员

快递

敬启者：

我方随函附上提交本合同争议仲裁意向书。我方今天已向合同另一方［补充雇主名字］提供了该通知及附函。合同依照［补充在封面上所显示合同名称］条款执行。贵方为被选定的提名者，因此，根据《建筑合同格式（英格兰与威尔士）条例 1998》中条款___和段落___的规定，我方谨此向贵方申请选定一人为仲裁员。随函附上已填妥的申请表格及相关方签发的总额［补充数额］的支票副本。

［如恰当，请补充：］

如果贵方不指定下述人员，我方将不胜感激：［列出在此案中指定机构小组中你不希望被提名的仲裁员名单，并简述理由］。

敬上

抄送至：雇主

Letter 3

To adjudicator, enclosing the referral

Special delivery

Dear

We note that you are/have been nominated as [*delete as appropriate*] the adjudicator. In accordance with clause ____ and paragraph ____ of the Scheme for Construction Contracts (England and Wales) Regulations 1998 we enclose our referral with this letter. Included are particulars of the dispute or difference, a summary of the contentions on which we rely, a statement of the relief or remedy sought and further material which we wish you to consider.

A copy of the referral and the accompanying documentation has been sent to [*insert name of employer*].

Yours faithfully

Copy: Employer with enclosures

[*The Referral must reach the adjudicator no later than 7 days after the date of the Notice of Adjudication.*]

信函 3

致仲裁者，附上介绍资料

快递

敬启者：

我方注意到贵方为仲裁者/已被指定为仲裁人［酌情删除］。根据条款____和《建筑合同格式（英格兰与威尔士）条例 1998》中段落____的要求，随函附上转介资料。资料包括纠纷或分歧详情，我方采信的争议总结，寻求救济或补救的声明以及我方希望贵方考虑的其他材料。

转介副本和所附文件已发给［插入雇主名字］。

敬上

抄送至：以附件形式发给雇主

［转介资料必须在裁决通知发出后七日内送达仲裁人。］

Letter 4

To employer, if the adjudicator's decision is in your favour

Special delivery

Dear

We have today received a copy of the adjudicator's decision. We note that the decision is in our favour. You will be aware that you must comply with the adjudicator's decision in accordance with the timescale laid down.

[*If appropriate, add:*]

Therefore, please [*insert the adjudicator's decisions and relevant time scales converted to actual dates, e.g., 'pay us the sum of* £40, 000.00 *by close of business on the* 10 *December* 2007].

If you fail to comply with the adjudicator's decision, we will immediately take enforcement proceedings through the courts, claiming interest, all our costs and expenses.

Yours faithfully

信函 4

致雇主，如果仲裁人的裁决对贵方有利

快递

敬启者：

今收到一份仲裁决定副本。我方知悉该决定对我方有利。请注意贵方必须按照规定的时间表遵守仲裁决定。

[如可行，则补充：]

因此，请 [将裁判员的决定和相关时间表转换为实际日期，例如，在 2007 年 12 月 10 日结束营业时，向我方支付 40，000 英镑的总金额]。

如贵方未遵守仲裁决定，我方将立即通过法院强制执行，同时追索利息，以及我方所有的花费。

敬上

Letter 5

To employer，requesting concurrence in the appointment of an arbitrator

Special delivery

Dear

We hereby give you notice that we require the undermentioned dispute or difference between us to be referred to arbitration in accordance with article ____ of the contract between us dated [*insert date*] . Please treat this as a request to concur in the appointment of an arbitrator.

The dispute or difference is [*insert brief description*] .

I propose the following three persons for your consideration and require your concurrence in the appointment within 14 days of the date of service of this letter，failing which we shall apply to the President or Vice-President of [*insert the name of the appointor as set out in the contract particulars or the abstract of particulars as appropriate*] .

[*List names and addresses of the three persons.*]

Yours faithfully

信函 5

致雇主，要求同意任命仲裁人

快递

敬启者：

谨此通知贵方，依据双方在［插入日期］签订的合同条款____规定，我方要求将下述双方的纠纷或分歧提交仲裁。我方请求贵方同意任命一名仲裁人。

纠纷或分歧是［插入简要描述］。

建议贵方考虑以下三人，请于收函起 14 日内同意任命，如不同意，我方将请求主席或副主席［插入合同详情中或详情摘要中列出的任命人名字］。

［列出三名建议人的名字和地址。］

敬上

Letter 6

To appointing body, if there is no concurrence in the appointment of an arbitrator

Dear

We are contractors that have entered into a building contract on [*insert the name of the contract as it appears on the cover*] terms, clause ____ of which makes provision for your President or Vice-President to appoint an arbitrator in default of agreement.

We should be pleased to receive the appropriate form of application and supporting documentation, together with a note of the current fees payable on application.

Yours faithfully

信函 6

致任命机构，如果不同意任命仲裁者

敬启者：

作为承包商，双方依据［插入合同封面上的合同名称］条款签订了一份建筑合同，该合同中条款____规定由贵方的主席或副主席任命一名合同违约仲裁员。

我发希望收到相关的申请表格和支持文件，以及申请时应付的费用。

敬上

Letter 7

To adjudicator, enclosing written statement

Special delivery

Dear

In accordance with your directions, we enclose a written statement setting out particulars of our response to the referral dated [*insert date*] and including the evidence on which we rely.

We are able to provide further information if you so require.

Yours faithfully

Copy: Sub-contractor

信函 7

致仲裁人，附上书面说明

快递

敬启者：

根据贵方指示，我方随函附上一份书面说明，详细列出对［插入日期］仲裁的答复，以及我方依据的证据。

如需要，我方可提供进一步信息。

敬上

抄送至：分包商

Letter 8

To sub-contractor if adjudicator appointed, but there is no dispute

Special delivery and fax

Dear

We note that you have sought the appointment of an adjudicator.

Your notice of intention to refer to adjudication contains no reference to a dispute or difference capable of being referred to adjudication. Therefore, the adjudicator has no jurisdiction. We invite you to withdraw and inform the adjudicator of the position. If you fail to do so, take this as notice that we will reserve all our rights and our participation in the purported adjudication will be without prejudice to our right to resist the enforcement of any decision.

Yours faithfully

Copy: Adjudicator

信函 8

致分包商，如指定仲裁人，但双方无异议

快递和传真

敬启者：

我方注意到贵方已经寻求任命一名仲裁人。

贵方的裁决意向通知书并未提及能够提交仲裁的争议或分歧。因此，仲裁人不具有管辖权。请贵方撤回，并将此情况通知裁决人。如果贵方不这样做，请视此函为我方所发出的通知，我方将保留我方所有权利，且我方参与的所谓裁决不会影响我方抵制执行任何决定的权利。

敬上

抄送至：仲裁人

Letter 9

To adjudicator, if there is no dispute

Special delivery and fax

Dear

The notice of intention to refer to adjudication which was submitted by the referring party on [*insert date*] contains no dispute capable of being referred to adjudication. Therefore, you are lacking jurisdiction and we invite you to relinquish your appointment.

If you fail to do so, take this as notice that we will reserve all our rights and our participation in the purported adjudication will be without prejudice to our right to resist the enforcement of any decision and to seek a declaration from the court that your decision is a nullity and, therefore, you are not entitled to fees.

A copy of the letter dated [*insert date*] which we have sent to the referring party is enclosed.

Yours faithfully

Copy: Sub-contractor

信函 9

致仲裁人，若无异议

快递和传真

敬启者：

相关方在［插入日期］提交的裁决意向通知书并未涉及可提交仲裁的争议。因此，贵方缺乏管辖权，请贵方撤销仲裁安排。

如果贵方不这样做，请视此函为我方所发出的通知，我方将保留我方的所有权利，且我方参加的所谓裁决不会影响我方抵制执行任何决定的权利。此外，我方将寻求法院宣判贵方决定无效。因此，贵方无权收取费用。

附上已函寄给相关方［插入日期］的信函副本。

敬上

抄送至：分包商

Letter 9 continued

[*Or*:]

The payment to which you refer was made on the [*insert date*].

[*Or*:]

An effective withholding notice was given on the [*insert date*].

[*Then add*:]

Your notice is therefore invalid and of no effect. Take notice that if you suspend further performance of your obligations under the sub-contract, such suspension will be unlawful. We reserve all our legal rights and remedies.

Yours faithfully

信函9 续:

[或者:]

贵方所提及的款项已于[插入日期]支付。

[或者:]

生效的扣缴通知书在[插入日期]已发出。

[然后补充:]

因此，贵方通知无效。望贵方注意，如贵方中止进一步履行分包合同规定的义务，此类中止将属违法行为。我方将保留所有法律权利和补救措施。

敬上

参　考　文　献

[1] 张水波，谢亚琴．国际工程管理英文信函写作．北京：中国建筑工业出版社，2001.

[2] Kit Werremeyer. Understanding & Negotiating Construction Contracts. Kingston，MA：RSMeans，2006.

[3] Garry Barnsley. Letters for Lawyers：Conveyancing. Australia：The Federation press，2000.

[4] Nael G. Bunni. The FIDIC Forms of Contract，(Third Edition). Blackwell Publishing，2005.

[5] ICE Conditions of Contract，Target Cost Version. First Edition，2006.

[6] Issaka Ndekugri，Michael Rycroft. The JCT Standard Building Contract Law and Administration (Second Edition). Elsevier，2009.

[7] AIA. Document A201—2007 General Conditions of the Contract for Construction.

[8] Joseph A. Huse. Understanding and Negotiating Turnkey and EPC Contracts (Second Edition). London：Sweet & Maxwell，2002.